Advances in Intelligent Systems and Computing

Volume 1071

The series "Advances in Intelligent Systems and Computing" contains publications on theory, applications, and design methods of Intelligent Systems and Intelligent Computing. Virtually all disciplines such as engineering, natural sciences, computer and information science, ICT, economics, business, e-commerce, environment, healthcare, life science are covered. The list of topics spans all the areas of modern intelligent systems and computing such as: computational intelligence, soft computing including neural networks, fuzzy systems, evolutionary computing and the fusion of these paradigms, social intelligence, ambient intelligence, computational neuroscience, artificial life, virtual worlds and society, cognitive science and systems, Perception and Vision, DNA and immune based systems, self-organizing and adaptive systems, e-Learning and teaching, human-centered and human-centric computing, recommender systems, intelligent control, robotics and mechatronics including human-machine teaming, knowledge-based paradigms, learning paradigms, machine ethics, intelligent data analysis, knowledge management, intelligent agents, intelligent decision making and support, intelligent network security, trust management, interactive entertainment, Web intelligence and multimedia.

The publications within "Advances in Intelligent Systems and Computing" are primarily proceedings of important conferences, symposia and congresses. They cover significant recent developments in the field, both of a foundational and applicable character. An important characteristic feature of the series is the short publication time and world-wide distribution. This permits a rapid and broad dissemination of research results.

** Indexing: The books of this series are submitted to ISI Proceedings, EI-Compendex, DBLP, SCOPUS, Google Scholar and Springerlink **

More information about this series at http://www.springer.com/series/11156

Jezreel Mejia · Mirna Muñoz ·
Álvaro Rocha · Jose A. Calvo-Manzano
Editors

Trends and Applications in Software Engineering

Proceedings of the 8th International
Conference on Software Process Improvement
(CIMPS 2019)

 Springer

Editors
Jezreel Mejia
Unidad Zacatecas
Centro de Investigación
en Matemáticas
Zacatecas, Zacatecas, Mexico

Mirna Muñoz
Unidad Zacatecas
Centro de Investigación
en Matemáticas A.C
Zacatecas, Zacatecas, Mexico

Álvaro Rocha
Departamento de
Engenharia Informática
Universidade de Coimbra
Coimbra, Portugal

Jose A. Calvo-Manzano
Escuela Técnica Superior de
Ingenieros Informáticos
Universidad Politécnica de Madrid
Madrid, Spain

ISSN 2194-5357 ISSN 2194-5365 (electronic)
Advances in Intelligent Systems and Computing
ISBN 978-3-030-33546-5 ISBN 978-3-030-33547-2 (eBook)
https://doi.org/10.1007/978-3-030-33547-2

This Springer imprint is published by the registered company Springer Nature Switzerland AG
The registered company address is: Gewerbestrasse 11, 6330 Cham, Switzerland

Introduction

This book contains a selection of papers accepted for presentation and discussion at the 2019 International Conference on Software Process Improvement (CIMPS'19). This conference had the support of the CIMAT A.C. (Mathematics Research Center/Centro de Investigación en Matemáticas), ROCKTECH, ITL (Instituto Tecnológico de León, Guanajuato, México), AISTI (Iberian Association for Information Sistems and Technologies/Associação Ibérica de Sistemas e Tecnologas de Informação), ReCIBE (Revista electrónica de Computación, Informática, Biomédica y Electrónica), VAZLO, and IBM. It took place at ITL, León, Guanajuato, México, from October 23 to 25, 2019.

The International Conference on Software Process Improvement (CIMPS) is a global forum for researchers and practitioners that present and discuss the most recent innovations, trends, results, experiences, and concerns in the several perspectives of software engineering with clear relationship but not limited to software processes, security in information and communication technology, and big data field. One of its main aims is to strengthen the drive toward a holistic symbiosis among academy, society, industry, government, and business community, promoting the creation of networks by disseminating the results of recent research in order to aligning their needs. CIMPS'19 is built on the successes of CIMPS'12; CIMPS'13; CIMPS '14; which took place on Zacatecas, Zac, Mexico; CIMPS'15, which took place on Mazatlán, Sinaloa; CIMPS'16, which took place on Aguascalientes, Aguascalientes, México; CIMPS'17, which took place again on Zacatecas, Zac, México; and the last edition CIMPS'18, which took place on Guadalajara, Jalisco, México.

The Program Committee of CIMPS'19 was composed of a multidisciplinary group of experts and those who are intimately concerned with software engineering and information systems and technologies. They have had the responsibility for evaluating, in a 'blind review' process, the papers received for each of the main themes proposed for the conference: Organizational Models, Standards and Methodologies, Knowledge Management, Software Systems, Applications and Tools, Information and Communication Technologies, and Processes in

Non-software Domains (mining, automotive, aerospace, business, health care, manufacturing, etc.) with a demonstrated relationship to software engineering challenges.

CIMPS'19 received contributions from several countries around the world. The papers accepted for presentation and discussion at the conference are published by Springer (this book), and the extended versions of best selected papers will be published in relevant journals, including SCI/SSCI and Scopus indexed journals.

We acknowledge all those who contributed to the staging of CIMPS'19 (authors, committees, and sponsors); their involvement and support are very much appreciated.

In loving memory of our International Scientific Committee members:

Rory V. O'Connor (1969–2019).
Luis A. Casillas Santillan (1971–2019).

October 2019   Jezreel Mejia
                                                                       Mirna Muñoz
                                                                      Álvaro Rocha
                                                                 Jose Calvo-Manzano

Organization

Conference

General Chairs

Jezreel Mejía   Mathematics Research Center, Research Unit
                      Zacatecas, Mexico
Mirna Muñoz   Mathematics Research Center, Research Unit
                      Zacatecas, Mexico

The general chairs and co-chair are researchers in computer science at the Research Center in Mathematics, Zacatecas, México. Their research field is software engineering, which focuses on process improvement, multimodel environment, project management, acquisition and outsourcing process, solicitation and supplier agreement development, agile methodologies, metrics, validation and verification, and information technology security. They have published several technical papers on acquisition process improvement, project management, TSPi, CMMI, and multimodel environment. They have been members of the team that has translated CMMI-DEV v1.2 and v1.3 to Spanish.

General Support

CIMPS General Support represents centers, organizations, or networks. These members collaborate with different European, Latin America, and North America organizations. The following people have been members of the CIMPS since its foundation for the last 8 years.

Gonzalo Cuevas Agustín	Politechnical University of Madrid, Spain
Jose A. Calvo-Manzano Villalón	Politechnical University of Madrid, Spain
Tomas San Feliu Gilabert	Politechnical University of Madrid, Spain
Álvaro Rocha	Universidade de Coimbra, Portugal

Local Committee

CIMPS established a Local Committee from the Mathematics Research Center; Research Unit Zacatecas, Mexico; ROCKTECH, Mexico; and the Technological Institute of León (ITL), Mexico. The list below comprises the Local Committee members.

Isaac Rodríguez Maldonado (Support)	CIMAT Unit Zacatecas, Mexico
Ana Patricia Montoya Méndez (Support)	CIMAT Unit Zacatecas, Mexico
Héctor Octavio Girón Bobadilla (Support)	CIMAT Unit Zacatecas, Mexico
Manuel de Jesús Peralta Márquez (Support)	CIMAT Unit Zacatecas, Mexico
Mario Alberto Negrete Rodríguez (Support)	CIMAT Unit Zacatecas, Mexico
Jorge Díaz Rodríguez (Support)	ROCKTECH, Mexico
Tania Torres Hernández (Support)	ROCKTECH, Mexico
Lic. Julián Ferrer Guerra (Local Staff)	ITL, Technological Institute of León, México
Rosario Baltazar Flores (Local Staff)	ITL, Technological Institute of León, México
M. C. Ruth Sáez de Nanclares Rodríguez (Local Staff)	ITL, Technological Institute of León, México
MIAC. Ana Columba Martínez Aguilar (Local Staff)	ITL, Technological Institute of León, México
MII. Margarita Sarabia Saldaña (Local Staff)	ITL, Technological Institute of León, México

Scientific Program Committee

CIMPS established an international committee of selected well-known experts in Software Engineering who are willing to be mentioned in the program and to review a set of papers each year. The list below comprises the Scientific Program Committee members.

Adriana Peña Pérez-Negrón	University of Guadalajara, Mexico
Alejandro Rodríguez González	Politechnical University of Madrid, Spain
Alejandra García Hernández	Autonomous University of Zacatecas, Mexico
Álvaro Rocha	Universidade de Coimbra, Portugal
Ángel M. García Pedrero	Politechnical University of Madrid, Spain
Antoni Lluis Mesquida Calafat	University of Islas Baleares, Spain
Antonio de Amescua Seco	University Carlos III of Madrid, Spain
Baltasar García Perez-Schofield	University of Vigo
Benjamín Ojeda Magaña	University of Guadalajara, Mexico
Carla Pacheco	Technological University of Mixteca, Oaxaca, Mexico
Carlos Alberto Fernández y Fernández	Technological University of Mixteca, Oaxaca, Mexico
Edgar Oswaldo Díaz	INEGI, Mexico
Eleazar Aguirre Anaya	National Politechnical Institute, Mexico
Fernando Moreira	University of Portucalense, Portugal
Francisco Jesus Rey Losada	University of Vigo, Spain
Gabriel A. García Mireles	University of Sonora, Mexico
Giner Alor Hernández	Technological University of Orizaba, Mexico
Gloria P. Gasca Hurtado	University of Medellin, Colombia
Graciela Lara López	University of Guadalajara, Mexico
Gonzalo Cuevas Agustín	Politechnical University of Madrid, Spain
Gonzalo Luzardo	Higher Polytechnic School of Litoral, Ecuador
Gustavo Illescas	National University of Central Buenos Aires Province, Argentina
Himer Ávila George	University of Guadalajara, Mexico
Hugo O. Alejandrez-Sánchez	National Center for Research and Technological Development, CENIDET, Mexico
Iván García Pacheco	Technological University of Mixteca, Oaxaca, Mexico
Jezreel Mejía Miranda	CIMAT Unit Zacatecas, Mexico
Jorge Luis García Alcaraz	Autonomous University of Juárez City, Mexico
José Alberto Benítez Andrades	University of Lion, Spain
Jose A. Calvo-Manzano Villalón	Politechnical University of Madrid, Spain
José Antonio Cervantes Álvarez	University of Guadalajara, Mexico
José Antonio Orizaga Trejo	University of Guadalajara CUCEA, Mexico

José Luis Sánchez Cervantes	Technological University of Orizaba, Mexico
Juan Manuel Toloza	National University of Central Buenos Aires Province, Argentina
Leopoldo Gómez Barba	University of Guadalajara, Mexico
Lohana Lema Moreta	University of the Holy Spirit, Ecuador
Luis Omar Colombo Mendoza	Technological University of Orizaba, Mexico
Magdalena Arcilla Cobián	National Distance Education University, Spain
Manuel Pérez Cota	University of Vigo, Spain
María de León Sigg	Autonomous University of Zacatecas, Mexico
María del Pilar Salas Zárate	Technological University of Orizaba, Mexico
Mario Andrés Paredes Valverde	University of Murcia, Spain
Mary Luz Sánchez-Gordón	Østfold University College, Norway
Miguel Ángel De la Torre Gómora	University of Guadalajara CUCEI, Mexico
Mirna Muñoz Mata	CIMAT Unit Zacatecas, Mexico
Omar S. Gómez	Higher Polytechnic School of Chimborazo, Ecuador
Patricia María Henríquez Coronel	University Eloy, Alfaro de Manabi, Ecuador
Perla Velasco-Elizondo	Autonomous University of Zacatecas, Mexico
Ramiro Goncalves	University Tras-os Montes, Portugal
Raúl Aguilar Vera	Autonomous University of Yucatán, Mexico
Ricardo Colomo Palacios	Østfold University College, Norway
Lisbeth Rodríguez Mazahua	Technological University of Orizaba, Mexico
Santiago Matalonga	University of the West, Scotland
Sergio Galván Cruz	Autonomous University of Aguascalientes, Mexico
Sodel Vázquez Reyes	Autonomous University of Zacatecas, Mexico
Sonia López Ruiz	University of Guadalajara, Mexico
Luz Sussy Bayona Ore	Autonomous University of Peru
Stewart Santos Arce	University of Guadalajara, Mexico
Tomas San Feliu Gilabert	Politechnical University of Madrid, Spain
Ulises Juárez Martínez	Technological University of Orizaba, Mexico
Vianca Vega	Catholic University of North Chile, Chile
Víctor Saquicela	University of Cuenca, Ecuador
Viviana Y. Rosales Morales	University of Veracruz, Mexico
Yadira Quiñonez	Autonomous University of Sinaloa, Mexico
Yilmaz Murat	Çankaya University, Turkey

Contents

Organizational Models, Standards and Methodologies

A Meta-Model for Regression Testing of Fintech Web Applications

Muhammad Saqlain Haider Malik[✉], Farooque Azam,
Muhammad Waseem Anwar, and Ayesha Kiran

EME College, NUST Islamabad, Islamabad, Pakistan
Muhammad.haider18@ce.ceme.edu.pk

Abstract. If a web application is migrated or the version of the system is changed then the test scripts becomes incapable of being validated because of the differences in the platform being used in testing. Regression tests need to be applied at that application after migration or version update are required to get written again in order to resolve compatibility issues with the new technologies. A problematic scenario occurs for the automation of script-level test when an application is updated or migrated just to include the technology which is not supported by the tool in use of testing. In this case, the purpose of test automation is eradicated because the scripts for regression testing are rewritten with the scripting language of the new testing tool being utilized. In this paper, a Model-Driven Approach is presented to develop the automated testing scripts for the validation of the FinTech Web Applications. A Meta-Model is proposed that depicts the scenario of regression testing of a FinTech system. A case study of small e-commerce platform known as Liberty Books is used for validation.

Keywords: Metamodel · Regression testing · FinTech regression · Web FinTech · Regression MetaModel

1 Introduction

A very costly and time-consuming endeavor is the Software testing. Software testing process consumes 50 to 75% the total development costs. The testing cost of the design and implementation tends to exceed thus the tools and methodologies used on these stages are very much important for the development of a quality software that is affordable as well. In reducing the time and costs of the testing, there is an effective way termed as the automation. So, a number of businesses conduct their software testing by the use of this automation. Test script level is typically the role to justify the level of automation for the software testing. A test-suite for any application is encoded by the software testers in a scripting language and then these encoded scripts are used as an input for an automated testing tool in the execution of tests. If multiple changes are made in the software, then the regression testing is performed automatically by the test scripts determining whether new errors are introduced in the previous version. A problematic scenario occurs for the automation of script-level test when an application is updated or migrated just to include the technology which is not supported by the tool in use of testing. In this case, the purpose of test automation is eradicated

J. Mejia et al. (Eds.): CIMPS 2019, AISC 1071, pp. 3–12, 2020.
https://doi.org/10.1007/978-3-030-33547-2_1

because the scripts for regression testing are rewritten with the scripting language of the new testing tool being utilized. To generate an executable code for any particular platform, the use of models and the model is emphasized by the MDSD (Model-Driven Software Development). The automated testing scripts generated for the validation of the FinTech Web Applications in this paper is done by using the MDSD. For the FinTech web based applications, a test set is to be designed in such a way that is not dependent on any platform and that is transformed further into a specific platform automated testing scripts automatically.

The term "FinTech" is a new word formulated from two words "Financial" and "technology" and it demonstrates the connection of the internet-related technologies i.e. cloud computing, internet banking, mobile internet with the business activities established by the financial services industry (e.g., transactions, billings). Innovators and the disruptors in the financial sector that uses the availability of ubiquitous communication, with the internet and the information processing automation are referred by the FinTech typically. The new tech seeking the improvement and automating the delivery along with the services of finances are described by the FinTech. The companies, owners of the business and the consumers manages their transactions/working of finance processes and the lives by utilizing specialized software algorithms utilizes the FinTech. All the things related to technology involves changes so the retesting is a must after certain changes. In history, the financial industries context took information technology as a tool for them. The established IT organizations and the FinTech start-ups those tend to enter the financial domain gain ground in the financial sector and customers are seized that are traditionally served by the established providers.

A safeguard procedure that verifies and validates the adapted system, and guarantee the purity from the errors is known as Regression Testing. The means to give surety regarding the required properties of the software once the changes occurred are provided by the Regression testing. Basically, this type of testing is performed for providing the confidence that the changes made do not affect the existing behavior of the software system. The test suite evolved by the software also grows in its size, resulting in increasing the cost to execute entire test suites. To justify the time and the relevant expenditure, regular regression testing is done with each and every update resulting in an array of additional benefits. Software testing is defined as firstly working with a software under pre-defined circumstances and after that maintaining/recording the results and then from these results an evaluation is made of some particular aspects of the system. Testing is also used in validating the changes when the maintenance is carried out in a software system. The test cases are rerun to determine whether there are any new errors introduced in the previously tested code is known as Regression Testing. A number of testing strategies take part with the automated tools in supporting the performance by the regression tests just as an effort to reduce the cost of testing.

The focus of MDSD is on the combination of the models of software and their related transformation in building a proper system. Typically, two things are involved in it i.e. use of the origin model that is the Platform independent model (PIM) and a targeted prototype that is the Platform specific model (PSM). PIM represents the essence of the solution because it does not rely on any specified technological platform. A language known as transformational language of model, is used to convert the PI-Model into a PS-Model which is then executed on the targeted platform. Meta-

Modeling is utilized in ensuring the consistency of the models/frameworks during transformations. The abstract syntax of the models and the inter-relationships between them are defined in this technique i.e. Meta-Modeling. Basically, this technique is a construction, analysis and the developments of the rules, frames, models, constraints and the theories that are applicable and useful in modeling a predefined problematic class.

For the designing and the development of automated testing scripts a model-driven approach is presented in this paper that validates the FinTech Web Applications. A Meta Model is provided for the regression testing of FinTech Web applications. The proposed modeling approach to support testing is validated through a case study i.e. Liberty Books.

2 Literature Review

The new technology seeking improvement and automating the delivery along with the use of financial services are described by using the FinTech. The companies, owner of the business and the consumers manages their working of finance transactions and the lives by utilizing specialized software algorithms utilizes the FinTech. All the things related to technology involves changes so the retesting is a must after certain changes. Regression Testing is known as a safeguard procedure that verifies and validates the adapted system, and guarantee the purity from the errors. Model Based Regression Testing (MBRT) has the ability to perform the tasks for tests in a good manner as the design models are used and the analysis for the changes are identified with the test cases in retesting the system after certain modifications.

A Meta model proposed [1] by the use of UML profiles describes that the trans-formations of models that are needed for doing the regression tests automatically. In [2], the three aspects of the FinTech systems are discussed as per the current situation. These aspects are the home and abroad development of the FinTech, requirement of the FinTech talents and their training status. The review on the requirement of providing a level of reliability of IS for the FinTech was provided [3]. On the basis of new communications and information technologies, digitalization is taken into considera-tion. In [4], evaluation is made on the state of the art in (MBRT) Model-based Regression testing with an examination of both the weaknesses and the strengths of the pre-defined approaches. Different approaches of model-based regression testing are also classified into Specification-based and design-based category. In [5], an approach for the FinTech black-box testing with a productive tool support for test oracles and the test generation both. The proposed approach is implemented on a CSTP subsystem. The possibility of happening a disruptive and collaboration between FinTech and the Bank is reviewed in [6] with the focus on peer to peer (P2P) mode. A new approach is proposed in [7] formulating a meta model that decreases the amount of the test cases to be taken into account for regression testing. An analysis is made in [8] that tends to provide an overview for the key value drivers behind FinTech that includes the use of Business models, Human centered design, resource-based theories and the open innovation.

In [9], focus is given on the observations and the views of managers along with testers and developers in organizations and stated that the test automation is costly but it improves the quality of the system. In [7], a survey is carried out to find out the gaps between the regression testing and the best practices in industry currently. A hybrid technique is proposed in [8] that carries a list of changes in source code, execution traces from the test cases that really combines modification, minimization and prioritization-based selection. In [10] it is proposed that meta-modeling has two forms of instantiation i.e. Linguistic and Ontological. A survey is conducted in [11] on the evaluation and comparison of the techniques of regression testing along with the techniques for web applications. In [12] an approach for the test selection of the Regression testing. Their approach utilizes the set of the system requirements that are usually written in the natural language and represents the part of the software that needs to get tested. A holistic approach is proposed in [13] to derive test cases for the regression testing from behavioral models that estimates the detection of the errors along with the test effort. Use cases are used in their approach for all estimates. In [14], an approach is proposed for prioritizing the test cases in model based environment. Activity diagram is used as it depicts the flow of each activity that is involved in the system. In [15], regression testing and the prioritization scheme is proposed as a model that selects the test cases from different user defined concerns for regression testing. In [16], an automatic tool is presented that is a java based tool used in reducing, generating and categorizing the test cases as reusable, obsolete and re-testable test cases. Model-driven approach [17] is presented for testing web applications. These strategies can be used utilized in designing a test for script generation.

Different types of testing are performed on the FinTech Systems but there isn't any specific research on the Regression Testing on the FinTech Systems. There are many models that works on the regression testing of the web applications but no work is done for the FinTechs. Certain models are made for the test selection and test prioritization of test suite but FinTech test cases are never referred. Detailed papers are written on the knowledge about the FinTech but regarding the testing of the FinTech systems there isn't any solid working. When changes are made to the system, it is required to perform regression testing with the update on previous version. Because of the growing technology, the FinTech practitioners have a little time to perform the regression test. Thus, a random testing approach is used or a little regression testing is performed. The lack of the required regression testing causes bugs in certain parts of the software system. Through effective testing methodologies the quality of the applications can be assured. Specific platforms for the creations of their models are used by maximum work focusing the usage of different models that helps in testing the web applications. In my paper, an approach is given to use a model that is platform independent and that helps out in the generation of test-scripts.

3   Proposed Solution

The dynamic integration of the services linked with both finance and technology are a referral to FinTech Systems. So, being critical systems it is necessary for FinTech Systems to be tested after any change as well i.e. regression testing. For automating the process of testing a FinTech System, it is required to develop a script that is understood by our testing tool, because after that it is applied with the firstly defined test cases to the system under tests. In FinTech systems, the services are interlinked with each other. Therefore, one change might affect several files/functionalities with it, but inadequate importance is given to regression test of this. To overcome this problem, a metal model is proposed for regression testing. All type, in which when any bug is fixed or code changes have been made, rather than just doing the regression test for that specific area, the regression test for all the affected area, and the related areas with it is done to ensure system accuracy & function. Another problem identified is inadequate importance to the test cases prioritization, whenever the regression test is done, test cases are neglected, because the test engineer focus is only the fixed/enhanced area, but test cases always ensure test coverage. The proposed meta-model also does prioritization of test cases for regression testing to ensure test coverage. Usually in FinTech systems, the systems are highly financial & trade with transactions worth millions depending upon the system. So making enhancements in the code sometimes compromises those securities, the end-to-end security of FinTech system while doing the regression testing is also considered. Another problem identified is the platform independency, it also does the test on multiple platforms to ensure platform independency.

3.1 Requirements for Proposed Solution

- Regression test Checklist to enlist the regression test areas.
- Test cases are required to cover the regression test to ensure high test coverage meaning high quality
- For platform independency & analytics, firebase is required to validate the performance & platform regression tests.
- Prioritization of the test cases.
- Checklist/report of the bugs resolved impacting which specific areas & the functional areas related to it.
- New functionality implemented affecting areas to be enlisted so that test engineer has an idea of the new functionality & it's affecting areas.

The use of a test script generator for producing a tests that are platform specific for the migrant/updated web systems is the key aspect proposed in this paper's methodology. The above mentioned model includes the inputs: Web Model 1 and Model 2 along with the Test Suite to the Test Script generator. As an output from the generator, the Test Sets i.e. Test set 1 and Test 2 are generated. Test Set 1 represents the Transactions and the accounts related tests while Test Set 2 represents the Web interface related test of the FinTech Web page. If the system migrates to a new set of version/technology, then test set automatically gets migrated by the combination of both the Test Set and the Web Model (Fig. 1).

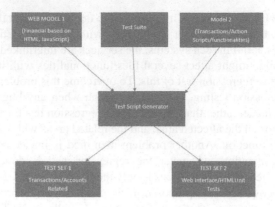

Fig. 1. Model-driven test script generation.

A conceptual model for the FinTech Web Applications is created that depicts the abstractions for the parts of the visual site/elements of the system that are important for the testing. It includes the FinTech Web Page and a Web Object of the Browser that firstly helps in hiding the complexities or problems of web Browser and then loads the FinTech Web Page on the Web Browser. Web Pages are basically static or interactive but using a FinTech system it needs to be an interactive webpage because it is a way of communication either giving order for some item or getting the payment done via different methods. Interactive Web Pages contains three sub items i.e. oracle form, transactions and Button Control respectively. Transactions are the regular processes and activities being carried out on the webpage. The Oracle Form is basically a link with the DB that is being managed. This Oracle Form contains the details regarding the Bank Details, Payment methods, Accounts and the billing reports that are linked to this form via bank details. The Web Browser is a formulation of a web object because a browser allows its user to go with the usual flow of control in an application so these type of scenarios are to be addressed by the testing. Browsers have their own configurations and the testing is set to be done with the browser's configuration's management. The compatibility of the Browser and the transactions on the FinTech webpage are then taken into account for testing and the test results are then maintained. Example of a scenario that shows the importance of the configuration management of the web Browser, a back button in a browser is pressed during the execution of application by the user causes unexpected results. In that case, Web browser facilitates changes management in the configuration of the browser and testing in allowed so that usage of a particular browser should be simulated (Fig. 2).

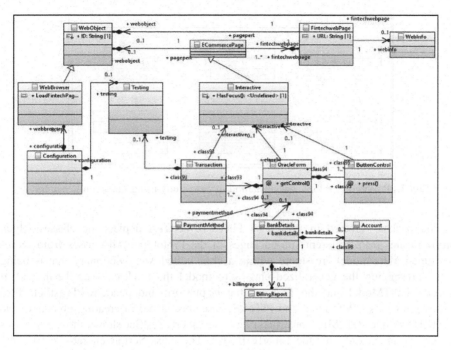

Fig. 2. Meta-model of a fintech web application

4 Case Study

We present a case study that is a proof showing the real concept of our methodology. Firstly, the features of the product/application are discussed along with the technologies and test support tools. After that details on the test set implementation is provided that is totally based on the proposed approach including the test script generation. Google Chrome and Mozilla Firefox were typically used to render this case study.

4.1 Liberty Books and Implementation

Liberty Books is an e-commerce application/platform for the purchase of books online. User is able to choose the category that what type of book is required and then the method of their billing is submitted along with all required information of shipping for validation of the user etc. This system has multiple categories including the categories, URDU books, reduced price and Gifts. Firstly, test cases for Liberty Books are formulated using the boundary value analysis (BVA) and equivalence partitioning techniques. The initially designed test set consist of certain test cases that are encoded with the constructs and the proposed Meta model.

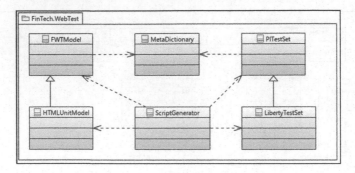

Fig. 3. Prototype for the implementation of case study using the script generator.

The package labeled in Fig. 3, as FinTech.WebTest depicts the classes which communicates from different sub-packages of the Prototype. All classes from aforementioned Meta-model are stored within a class called MetaDictionary that is being used in designing the generalized classes to model the FinTech Web Testing technologies FWTModel and the Test set that is platform independent PITestSet. Two classes known as FWTModel and PITestSet are specialized in creating objects of the HTMLUnitModel (HTML Constructs) and the LibertyTestSet shows those test cases that are designed only for the Liberty Books. The class ScriptGenerator carries the responsibility to iterate among the LibertyTestSet, and the platform specific HTML constructs in the HTMLUnitModel are used in generating test scripts. These Scripts are achieved by the retrieval of the definition of the test case that is representing the abstract test commands, along with the variable title and the related values. The represented abstract test commands are further stored into the HTMLUnit constructs along with the placeholders for variable titles and values. After that, the placeholder is overwritten by generator with the original name with values that are then stored in the LibertyTestSet and the complete instructions are appended into the Liberty.HTMLUnit. An instance diagram is also made to validate the case study (Fig. 4).

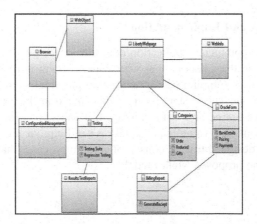

Fig. 4. Object diagram for the validation of case study

The web based system i.e. Liberty Books is opened on a Browser, browser has a web object that helps in displaying the web page and the web page is consisting of the related web info. Liberty WebPage is an interactive web page having various transactions based on the categories along with the Banking details that make it's a financial system. Categories that are included in Liberty Books are the Urdu Books, Reduced Price and Gifts. Oracle form is having the Pricing method details that has the banking detail and the billing report that further generates a receipt. Testing is carried out on the transactions that includes the order along with the Bank related payment details and the configuration of the browser that stores results of testing in Test Report object.

5   Discussion

In this study, the demonstration of the application of a model-driven approach in designing and developing the scripts for testing FinTech web applications is done. It is able be used for defining the tests to be Platform independent though converting them into the scripts that are perfect to be used for an automated testing tool. The deep study helps out in having a clear view in developing a framework for the proposed approach. The generation of the automated test scripts are modelled by the test case that addresses the threat connected with the technological changes and migration among web applications. System particularly an e-commerce for books i.e. Liberty Books is taken into account as our case study and a. There are some limitations of the proposed prototype. **One of the limitation is the coverage of a subset of web controls and the manual detection of constraint violations by model. This limitation can be latter resolved by using the tools of the model-driven architecture i.e. Eclipse Modeling Framework.**

6   Conclusion and Future Work

In this paper, a model-driven approach is presented for the designing and the development of automated testing scripts that validates the FinTech Web Applications. In FinTech systems, the services are interlinked with each other. Therefore, one change affects several files/functionalities with it, but inadequate importance is given to regression test of this. Therefore, a generic meta-model is proposed for the regression testing of FinTech Web applications. The proposed solution can be extended so that it includes the technology model for new technologies as well as FinTech are not comprised with just the simple Technology. A similar Meta-model can also be developed for the FinTech mobile applications as well. **The Technological model for flash and Actionscript will be used to extend the prototype. Additional test scenarios will be formulated for the Liberty Books.**

References

1. Hernandez, Y., King, T.M., Pava, J., Clarke, P.J.: A Meta-model to support regression testing of web applications. School of Computing and Information Sciences, Florida International University Miami, FL 33199
2. Mei, S., Peiguang, L. Xiushan, N.: Research on fintech industry development and talent training status, Jinan, Shandong Province, China
3. Pantielieieva, N., Krynytsia, S., Khutorna, M., Potapenko, L.: FinTech, transformation of financial intermediation and financial stability
4. Farooq, Q., Riebisch, M., Lehnert, S.: Model-based regression testing—process, challenges and approaches. Ilmenau University of Technology, Germany
5. Wang, Q., Gu, L., Xue, M., Xu1, L., Niu, W., Dou, L., He, L., Xie, T.: FACTS: automated black-box testing of FinTech systems
6. Saifan, A.A., Akour, M., Alazzam, I., Hanandeh, F.: Regression test-selection technique using component model based modification: code to test traceability
7. Lin, X.: Regression testing in research and practice. Computer Science and Engineering Department, University of Nebraska, Lincoln 1-402-472-4058
8. Wong, W.E., Horgan, J.R., London, S., Agrawal, H.: A study of effective regression testing in practice. Bell Communications Research, 445 South Street Morristown, NJ, 07960
9. Taipale, O., Kasurinen, J., Karhu, K., Smolander, K.: Trade-off between automated and manual software testing. Int. J. Syst. Assur. Eng. Manage. 2, 114–125 (2011)
10. Atkinson, C., Kuhne, T.: Model-driven development: a metamodeling foundation. IEEE Softw. 20(5), 36–41 (2003)
11. Hassnain, M., Abbas, A.: A literature survey on evaluation and comparison of regression testing techniques for web applications
12. Chittimalli, P.K., Harrold, M.J.: Regression test selection on system requirements
13. Prabhakar, K., Rao, A.A., Rao, K.V.G., Reddy, S.S., Gopichand, M.: Cost Effective Model Based Regression Testing. In: Proceedings of the World Congress on Engineering, vol. 1 (2017)
14. Shinde, S.M., Jadhav, M.K.S., Thanki, M.H.: Prioritization of re-executable test cases of activity diagram in regression testing using model based environment. 3(7). ISSN 2321-8169
15. Silva Filho, R.S., Budnik, C.J., Hasling, W.M., McKenna, M., Subramanyan, R.: Supporting concern based regression testing and prioritization in a model-driven environment. In: 34th Annual IEEE Computer Software and Applications Conference Workshops (2010)
16. Devikar, R.N., Laddha, M.D.: Automation of model-based regression testing. Int. J. Sci. Res. Publ. 2(12) (2012)
17. Heckel, R., Lohmann, M.: Towards model-driven testing. Electr. Notes Theor. Comput. Sci. 82(6), 33–43 (2003)

Objectives Patterns Applied to the Business Model of a Public Education System

Brenda Elizabeth Meloni[1], Manuel Pérez Cota[2],
Oscar Carlos Medina[1(✉)], and Marcelo Martín Marciszack[1]

[1] Facultad Regional Córdoba, Universidad Tecnológica Nacional,
Cruz Roja Argentina y Maestro López s/n, Ciudad Universitaria,
Córdoba, Argentina
bemeloni@gmail.com, oscarcmedina@gmail.com,
marciszack@frc.utn.edu.ar
[2] Universidad de Vigo, Campus Universitario s/n, 36310 Vigo,
Pontevedra, Spain
mpcota@uvigo.es

Abstract. *Context:* A pattern is a description of a common solution to a recurring problem that can be applied in a specific context. There are different kinds of patterns in Software Engineering, such as Business Patterns. *Goal:* The present work develops a particular kind of Business Patterns, called Objectives Patterns, used to model business processes. The application of these patterns to a Business Model of a Public Education system is also analysed. *Methods:* The process that supports the Education System is analysed and Objectives Patterns are applied, according to the proposal stated in the bibliography, which was reviewed in the Theoretical Framework. *Conclusions:* The application of Objectives Patterns allowed elucidating the requirements, analysing possible difficulties, and finding solutions and similarities between scenarios to simplify the process modelling.

Keywords: Objectives Patterns · Business Model · Education System · Electronic Government

1 Introduction

It is feasible to find many cases of software products that misrepresent or incorrectly support the Business Model which they intend to solve. Some of the causes for this situation are an incomplete requirements elicitation, ignorance of the business difficulties, incorrect modelling and others.

The Business Model allows an appropriate business abstraction, knowing its objectives, processes, activities, resources and rules, especially in complex businesses. In addition, the comprehension of the Business Model allows to speak the same language as the client, which is useful for negotiation and to overcome ambiguities, to limit the scope of the system to develop and to validate the software deployed. One of the business modelling techniques is through Business Modelling Patterns.

© Springer Nature Switzerland AG 2020
J. Mejia et al. (Eds.): CIMPS 2019, AISC 1071, pp. 13–22, 2020.
https://doi.org/10.1007/978-3-030-33547-2_2

A pattern is a description of a common solution to a recurring problem that may be applied in a specific context [1]. Patterns help to take advantage of the collective experiences of specialized software, where engineers represent an existing proven experience in systems development. Patterns help to promote good design practices.

According to Eriksson y Penker [2], each pattern handles a determined recurring problem in the design or implementation of a software system. Patterns can be used to build software architectures with specific properties. These authors state that Business Modelling Patterns allow to obtain the resources from the model, organizations, objectives, processes and rules of the company. They classify them in three categories: Resources and Rules Patterns, Processes Patterns and Goal or Objective Patterns.

Here we present the development of Objective Patterns to model business processes and the analysis of its application on the Business Modelling of a Public Education System that supports the Ministry of Education of Córdoba Province, Argentina.

2   Objective Patterns Theoretical Framework

The patterns in a Business Model are called Business Patterns, with Goals or Objectives Patterns classes as a sub-type. The main concepts to understand the representation of these relationships will be explained further on.

2.1   Business Model

The business process, supported by the information system, is based on the Business Model, which is defined as: "*a conceptual tool that contains a set of objects, concepts and their relationships with the objective to express the business logic of a company*" [3]. Al Debei, El Haddadeh and Avison [4] define it as "*an abstract representation of an organization, be it conceptual, textual, and/or graphical, of all arrangements that are needed to achieve its strategic goals and objectives*".

The Business Modelling allows representing an organization, to obtain the processes that form it, to know the structure and to understand the business logic, rules and actors.

2.2   Pattern Concept

According to Alexander [5], a pattern is a "*generalized solution that may be implemented and applied in a difficult situation, thus removing one or more of the inherent problems in order to meet one or more goals*". He was the architect that proposed the pattern concept in his book, "The Timeless Way of Building", in 1979.

Based on Alexander studies, Beck and Cunningham developed a five-pattern language for Smaltalk beginner programmers [6]. Coplien then built a patterns catalogue for C++ and published it in 1991 [7].

Later, Gamma, Helm, Johnson, and Vissides, published the most important work on patterns applied to software design: "Design Patterns: Elements of Reusable Object-Oriented Software" [8], in concordance to the object oriented programming paradigm.

Patterns are now being used to describe solutions to problems of Software Engineering, as a guide for the different stages of the construction of a system, creating specific types for each one of them, such as the case of patterns that model business processes.

2.3 Objectives Patterns

There are different types of patterns in Software Engineering, according to the characterization made by Sommerville [9]. The following are some of the best known patterns:

- Patterns found in a Business Model, called Business Patterns.
- Patterns in the design of a system are known as Architectural Patterns.
- Patterns closer to the programming level are Design Patterns.

In their book, "Business Modelling with UML: Business Patterns at Work" [2], Eriksson y Penker, present three categories of Business Patterns:

- Processes Patterns.
- Resources and Rules Patterns.
- Objectives and Goals Patterns.

Processes Patterns are behaviour patterns which intend to improve the quality of the processes, workflows and other models.

Resources and Rules Patterns provide guidelines for the modelling of rules and resources inside de business domain.

Objectives or Goals Patterns are models that describe the problem-solution relationship with the objectives of a business process. The modelling of objectives is a subject considered critical since it affects in a global manner the system to be built. This approach allows to infer that processes describe the way to meet the specified goals with a given set of resources and rules. There are previous publications [1, 10–13] of the authors of this work that develop an analysis model for the application of patterns.

Following this line of research, the use of objective patterns to describe Electronic Government processes will be now verified.

3 Materials and Methods

In this section we will present the application of Objectives Patterns in the Business Modelling of an academic management system of Córdoba Province Government: the Province Education System. The business process in which this software is bases is complex, with many rules and exceptions. The case study in this work has defined one of the business process of this system as scope, and Objective Patterns where applied to it.

3.1 Selected Study Case

The Education System of Córdoba Province is a big and complex system that includes approximately 5000 public and private schools, and around 900.000 students, within four educational levels and four types of education (see Table 1). The Ministry of

Education is formed by managements in accordance to combination of levels and types of education, which is the academic offer provided by the province.

Table 1. Relationship between academic levels and types in the Educational System.

Levels	Academic types			
	Common	Special	Adults	Technical
Initial	X	X		
Elementary	X	X	X	
Mid	X		X	X
High	X			X

The academic levels are related to a group of the student population of Córdoba Province, and are adapted to the group. Therefore, the rules between each of them have a big variance, in the same manner as the education types and the combination between them.

3.2 Business Process: Enrolment to a School or Institute

Enrolment to a school or institute is the process of registering a person as a student in a school or institute. This definition is applied to every academic level, with some exceptions. In higher education, a person is enrolled once in an institution and career, while in the other levels students must register in every academic year. Córdoba Province only allows a student to enrol in a single school, while in the higher level a student can register in many schools since he can study more than one career and they can be in different institutions.

Figure 1 presents the general Enrolment process:

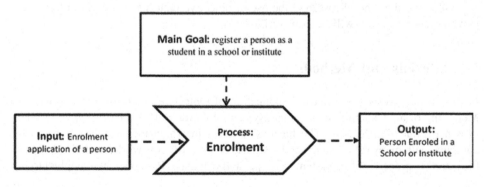

Fig. 1. Enrolment process diagram.

3.3   Objective Pattern Applied to Business Process: Enrolment to a School or Institute

Goal or Objectives Patterns are used for objectives modelling. It is a very critical issue since it affects the entire process on how to model the system that will be built. Process models refer to resources and are restricted by rules in order to meet the goals of the process. It can be said that the processes describe how to achieve the specified objectives with a set of predefined resources and rules.

Process Patterns are behaviour patterns whose intention is to increase the quality of processes, workflows and other models.

Resource and Rules Patterns provide guidelines for modelling rules and resources within the business domain.

As it was present in the Theoretical Framework, the Objectives Pattern exists to model objectives in business processes. It is formed by three other Patterns: Objective Breakdown Pattern, Objective Assignment Pattern and Problem – Objective Pattern.

The technique used is the one proposed by the authors of this work in a recent publication [1], where the results of the last research project are gathered. The objective of that project was the implementation of patterns to validate conceptual models. Considering the variety of academic level and education types presented in Table 1, we selected the Elementary level with a common education type in order to apply the patterns.

3.4   Objectives Patterns Application

Every elementary student with a common educational type is enrolled to a grade and division, for instance, to second grade D. Student personal data is also registered, as well as the data from their parents or tutors.

The first step is to apply de Objective Breakdown Pattern, then the Objective Assignment Pattern and finally de Objective – Problem Pattern.

3.4.1   Objective Breakdown Pattern

The following pattern can be obtained starting from the global objective and from the inherent features of each academic level and type of education:

Global Objective: enrol a person, a student in a school or institute.

- Sub-Goal A: register a person in a grade/division of the school.
  - Sub-goal A1: register a person of the school.
  - Sub-goal A2: register a person from another school.
  - Sub-goal A3: register a person to the Academic Level for the first time.
- Sub-goal B: register the personal data of the student from the documentation received.
- Sub-goal C: register the personal data of the parents or tutors.

3.4.2   Objective Assignment Pattern

The Enrolment process objective is defined globally to be able to cover every level of the Educational System. The Objective Assignment Pattern allows identifying sub-processes to cover every business activity, rule and resource involved to meet the global objective and, therefore, gain a better understanding of the Business Model.

According to the application of the Objective Assignment Pattern three sub-processes can be identified:

- Sub-Goal A: Enrolment Management Sub-Process
- Sub-Goal B: Personal Data Management Sub-Process
- Sub-Goal C: Guardian Management Sub-Process.

3.4.3   Objective – Problem Pattern

We found different difficult situation when applying this pattern in the stated objectives. The following application example in Sub-Goal A3 is presented in Fig. 2:

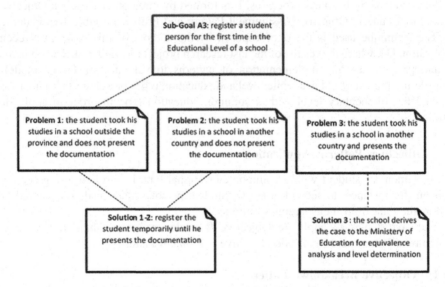

Fig. 2. Objective – Problem Pattern applied to Sub-Goal A3.

4   Results

Objective Patterns were applied in a previous study over the adult education level, in the same Enrolment to School and Institute Business Process, and the Pattern could be applied to the following levels and types of education, called Objectives Pattern A (OP-A):

- Mid-Level with adult education type – on – site course
- Mid-Level with adult education type – distance course
- Higher Level with technical education type.
- Higher Level with common education type.

In the practical case explained in the present work the study is carried on over the Business Model and in the same Process, obtaining an Objective Pattern in the elementary level with a common education type. The combination of remaining levels and education types in the Educational System will be analysed to verify if the generated pattern can be reused in the following section.

(a) Mid-Level with common and technical education type.

In this level, students are enrolled to a single school, in an academic year or division and inside a curriculum. The grade or division of the Elementary level is replaced by year or division. The Sub-Goals of Objective A also apply to this level, as well as the Sub-Processes Enrolment Management, Personal Data Management and Guardian Management that emerge from them.

(b) Initial Level with Common education type.

Initial level is formed by classroom instead of grades or years. The students enrol to a single school, to a classroom or division and to a curriculum. Enrolment Management, Personal Data Management and Guardian Management Sub-Processes also apply.

Thus, the Objective Pattern obtained for the elementary level with common education type is applicable to the four level and education type combinations of the Educational System of Córdoba Province, and we will call it Objectives Pattern B (OP-B):

- Elementary Level with common education type
- Initial Level with common education type
- Mid-Level with common education type
- Mid-Level with technical education type.

Objective Patterns were applied to a same Academic Level with two types of education and, despite de differences in rules, methods and conditions, it was possible to obtain two different Objective Patterns that cover every level and type of education in the Educational System. The best-case scenario would have been to be able to have a single Pattern, but there are more exceptions than rules in educations. Being able to reduce so many difficulties in two simple Patterns is a proof of the advantages of using these types of Patterns for the Business Models analysis.

The Objective Patterns were applied to the same Educational Level with two types of education and despite the differences in rules, modalities and conditions, it has been possible to obtain two different Objective Patterns that encompass all levels and type of education of the Educational System.

The optimum would have been to have only one employer, but in this case the aggregate is required for exception control. The achievement of simplifying this complex casuistry in two simple patterns, OP-A and OP-B, shows the advantages of using these types of patterns for the Business Models analysis (see Table 2).

The application of Objective Patterns was performed in the order recommended by the bibliography: Objective Breakdown Pattern, Objective Assignment Pattern and Objective Problem – Pattern. Other types of processes have, however, been identified, and it is needed to re-apply patterns to them in order to assign objectives and complete the Objectives and Sub-Goals of the process. In addition, every new process obtained

Table 2. Objective Patterns associated with the level directions, educational levels and types of education of the Educational System.

Level directorate	Levels	Academic types	Pattern
General directorate of youth and adult education	Elementary	Adults	OP-A
	Mid		
General directorate of secondary education	Mid	Common	OP-B
General directorate of technical education and vocational training	Mid	Technical	OP-B
	High		OP-A
General directorate of higher education	Initial	Common	OP-B
	Elementary		
	Mid		
	High		OP-A
General directorate of primary education	Initial	Common	OP-B
	Elementary		
General directorate of initial education	Initial	Common	
General directorate of private education	Initial	Common	
	Elementary		
	Mid	Common	
		Technical	
	High	Common	OP-A

must be dealt with the Objectives Patterns, finding difficulties and solutions, and a new process may emerge, and so on, until a fine grain is reached for the activities of each process.

If we could go deeper to the lowest level of specification through these patterns, there would be no doubts, no ambiguities, and no lack of information or understanding in any analysed process. The highest grade of detail for the pattern is considered to be reached at that instance.

5   Conclusions

The understanding of the Business Model must be as detailed as possible during requirements elicitation and in each Business Process that forms it, in order to reach a full abstraction. This is especially true in complex models such as the one present in the Educational System of Córdoba Province.

Objective Patterns are truly important to know the activities of a Business Process and to discover other instances that are not as clear to the people responsible for the processes. This is done by breaking down each objective to the lowest level and laying out the difficulties. The application of these patterns in order to obtain a successful modelling of objectives, and doing it at the beginning of the Business Model analysis, speeds up the study of the processes, especially in complex systems such as the one analysed in this work, which presents different complex and variable scenarios.

The application of the Objectives Pattern allowed to clarify the requirements elicitation, to lay out difficulties, and to find solutions and similarities between the scenarios in order to simplify the process modelling.

These patterns show bigger advantages when they are introduced at the start of the analysis, to agree with the person in charge of the process objectives. Often, this early activity does not get the needed importance and its impact in maintainability is not considered for the software product built.

It is concluded in the first place that the Patterns of this article should be included in a E-Gob patterns catalogue together with the Patterns developed in previous works [12, 13]. And they can be used in the experimentation stage of an investigation that aims to define an analysis model for the application of Patterns in the Conceptual Modelling of Electronic Government systems [11].

We propose to apply this technique at the beginning of the analysis to elucidate the Business Model and visualize its structure more clearly, especially in complex processes with different combinations of courses of action. The introduction of patterns is not very time consuming, in comparison to the valuable information obtained. In addition, it does not require a higher complexity to be able to use them in daily analysis processes. It is a good practice to recommend in business processes studies.

The representation of patterns through graphics allows to have a better vision of the business with its Objectives and Sub-Goals, allowing to easily identify the process and its activities, and to be able to validate the reuse of the pattern. It also allows to have an accurate and common vocabulary between the person responsible for a process and its analyst, removing ambiguities and refining requisites.

References

1. Marciszack, M.M., Moreno, J.C., Sánchez, C.E., Medina, O.C., Delgado, A.F., Castro, C.S.: Patrones en la construcción del Modelo Conceptual para sistemas de información. Editorial edUTecNe, U.T.N (2018)
2. Eriksson, H.-E., Penker, M.: Business Modeling with UML: Business Patterns at Work. OMG Press, Wheat Ridge (2000)
3. Osterwalder, A., Pigneur, Y.: Generación de Modelos de Negocio, 3rd edn. Wiley, Hoboken (2011)
4. Al-Debei, M.M., El-Haddadeh, R., Avison, D.: Defining the business model in the new world of digital business. In: Proceedings of the Americas Conference on Information Systems (AMCIS), vol. 2008, pp. 1–11 (2008)
5. Alexander, C.: The Timeless Way of Building. Oxford University Press, New York (1979)
6. Beck, K., Cunningham, W.: Using pattern languages for object-oriented programs. Submitted to the OOPSLA-87 workshop on the Specification and Design for Object-Oriented Programming (1987)
7. Coplien, J.: Advanced C ++ Programming Styles and Idioms. Addison Wesley, Boston (1992)
8. Gamma, E., Helm, R., Johnson, R., Vissides, J.: Design Patterns: Elements of Reusable Object-Oriented Software. Addison-Wesley, Boston (1994)
9. Sommerville, I.: Ingeniería de Software, 9th edn. Pearson, London (2011)

10. Medina, O.C, Marciszack, M.M., Groppo, M.A.: Un modelo de análisis para aplicación de patrones de buenas prácticas en el modelado conceptual de gobierno electrónico. In: Proceedings of WICC 2018. Red UNCI y UNNE Universidad Nacional del Nordeste (2018)
11. Medina, O.C, Marciszack, M.M., Groppo, M.A.: Proposal for the patterns definition based on good practices for the electronic government systems development. In: Proceedings of CISTI 2018 – 13th Iberian Conference on Information Systems and Technologies (2018)
12. Medina, O.C, Cota, M.P., Marciszack, M.M., Martin, S.M., Pérez, N., Dean, D.D.: Conceptual modelling of a mobile app for occupational safety using process and objectives patterns. In: Trends and Applications in Software Engineering, pp. 186–195. Springer (2018)
13. Medina, O.C, Cánepa, P.A., Gruppo, M.O., Groppo, M.A.: Un caso de estudio de patrones de Gobierno Electrónico para gestión de consultas de ciudadanos. In: proceedings of CONAIISI 2018, Red RIISIC, CONFEDI y Universidad CAECE (2018)

Requirements Validation in the Information Systems Software Development: An Empirical Evaluation of Its Benefits for a Public Institution in Lima

Luis Canchari[1] and Abraham Dávila[2]($\boxtimes$)

[1] Facultad de Ingeniería, Universidad Continental, Huancayo, Perú
lcanchari@continental.edu.pe
[2] Departamento de Ingeniería, Pontificia Universidad Católica del Perú,
Lima, Peru
abraham.davila@pucp.edu.pe

Abstract. Under Peruvian legislation, all information technology offices belonging to public institutions must meet ISO/IEC 12207 and, in the case of small organizations the international standard series ISO/IEC 29110 could be a good strategy to comply with the legislation. In this context, ISO/IEC 29110-5-1-2 establishes to apply software validation to requirements as a relevant activity; but it is an unusual practice during the software development of information systems in Peru. In this study, we assess the benefits of software validation in the context of a public institution technical unit. For this purpose, a case study was performed and a quasi-experimental design was applied on a group of projects. The quality indicators defined increased their values as a consequence of process improvement based on software validation. As a conclusion, applying software validation to requirements is beneficial and contributes to increase project quality.

Keywords: Software requirements · Validation of software requirements · Information systems

1 Introduction

The Peruvian government, in 2004, established the mandatory use of the national standard equivalent to ISO/IEC 12207 [1]. Thus, this legal framework forces information technology units (ITU) to take actions for their adoption and auditors to verify compliance [1]. However, almost all of these ITUs are very small and they usually take care of software development and software operation tasks (installation and help desk); so the effective number of people dedicated to software development is lower (frequently less than 25 professionals). On the other hand, in the software industry many companies are small and have particular characteristics [2, 3] and according to some authors, implementing large models such as ISO/IEC 12207 or CMMI is complicated [4, 5] for them. Therefore, this situation forced to develop models oriented to small companies as MoProSoft [6], MPS-Br [7] and Ágil SPI [8], among others. In particular,

© Springer Nature Switzerland AG 2020
J. Mejia et al. (Eds.): CIMPS 2019, AISC 1071, pp. 23–35, 2020.
https://doi.org/10.1007/978-3-030-33547-2_3

in 2011, ISO/IEC 29110-5-1-2 was published, a guideline for small software development organizations [4]. This guideline is based on ISO/IEC 12207 and ISO/IEC 15289 [3]. Considering the above, it can be noted that ISO/IEC 29110-5-1-2 becomes a good strategy to initiate the adoption of ISO/IEC 12207 in small organizations. Under national standardization office there are NTP-ISO/IEC 12207 and NTP- ISO/IEC RT 29110-5-1-2 as a translation to Spanish of ISO/IEC 12207 and ISO/IEC TR 29110-5-1-2 respectively [9].

The ISO/IEC 29110-5-1-2 is a guideline for the implementation of the basic profile for software engineering, and it includes verification and validation tasks; in particular, one of them establishes software validation of requirements [10]. In this sense, it should be noted that different studies have tried to establish the relationship between software requirements and their success [11], others noted that the requirements analysis is one of the most important phases [12–15]. Also [16] points out that requirements issues have originated the software crisis. All of the above shows that the quality actions oriented to requirements should contribute to the project performance and to the software product quality obtained.

In our experience in different ITU, based on ten years of interaction with ITUs, in the context of public institutions in Lima, it is known that due to the pressure to deliver products in the short term, software developers of the ITU leave aside activities that do not generate tangible deliverables for the users. This means that in most cases they leave out activities associated with verification and validation, among others. This breach, as pointed out by different authors [17–20], causes an increase in the number of defects, some are detected during development and others in production, which results in negative impacts on scope, time, quality and image in front of users.

In this article we present an assessment of software requirements validation in terms of its benefits in the context of the ITU, which develops information systems software for a public institution in Lima. In Sect. 2, a framework is presented which includes some key concepts and the state of art; in Sect. 3 the case study design is presented; in Sect. 4, the results are shown; and in Sect. 5, the final discussion and future work is presented.

2   Background

In this section, some basic concepts are presented and relevant works related to our study.

2.1   Basic Concepts

The key concepts for this article are:

Software Requirements. According to ISO/IEC 24765 [21], there are two definitions: (i) a software capability needed by a user to solve a problem or to achieve an objective; and, (ii) a software capability that a system or system component must meet or have to satisfy a contract, standard, specification or other formally imposed document. Some authors of software engineering textbooks [12, 22], point out that the requirements are declarations, which in addition to fulfilling the above, can be high level or detailed, functional or non-functional; among other considerations.

Software Validation. According to ISO/IEC 12207 [23], it is the confirmation, through objective evidence, that the requirement for a specific expected use or application has been met. It also points out that during the development process; the validation implies a set of activities to gain confidence that the software will meet the established requirements.

Software Requirements Validation. As indicated in the previous items, it is the confirmation that the requirements are expressed in an appropriate manner and corresponding to the user needs. According to [24, 25], various actions can be carried out during the process of collecting requirements to ensure that they fulfill the user needs and are the basis for building the required software.

2.2 Related Work

The validation techniques of software requirements are, in general, procedures and resources to ensure that the requirements reflect the user needs and that can be built. In [26], as part of the article, it presents a literature mapping where the two most used techniques are identified: previous reviews and requirements reading [26], other techniques are identified as: inspections based on test cases (TDC, test case driven), requirement inspection, prototyping and validation based on model. On the other hand, [27] mentions that the techniques should be contextualized, and adjusted if necessary, depending on the organizational culture, experience and skills of the people involved.

In the works related to this article, it can be mentioned that: (i) in [24] a methodology is proposed for the acquisition and management of software requirements for small and medium enterprises, which takes elements and practices from CMMI, SWEBOK, ITIL, Competisoft and MPIu + a; (ii) in [28] a procedure is proposed to specify and validate software requirements in MIPYMES based on prototyping and walkthroughs that allowed to increase the detection of defects; (iii) in [16] an iterative process model based on elicitation, analysis and validation of requirements is proposed; (iv) in [29] it is pointed out that an important aspect is the detection of defects, therefore it proposes a model based on the inspections technique; and, (v) in [30] automatic techniques are developed which support the process of development, validation and verification of software requirements, based on the automatic analysis techniques coming from formal methods.

Other related works with more information for the final discussion are:

In [31] it is presented: (i) a study about the requirements impact on software quality; (ii) the results of applying static tests to a requirements sample in a financial institution, showing that no document fully complies with the quality characteristics and only 30% meet some characteristics of a good requirement; (iii) most of the requirements assessed do not meet the feasibility, correct, complete and unambiguous characteristic; however, it is showed that the aspect in which the requirements have an optimal compliance is the consistency; and (iv) it was verified that for the projects under testing process, 23% of the total defects reported were due to inadequate requirements definition.

In [26] it is presented: (i) the financial impact of the software validation adoption in the analysis phase of small projects in a government organization; (ii) two projects are compared, one of them incorporated the requirements validation through review, which allowed taking corrective actions in the scope; (iii) when controlling the number of functional errors and the impact on change requests, it was observed that the cost of these was reduced from 16.9% to 10.1% of the initial budget; (iv) it is concluded that the software requirements validation introduction in the project was a success and the organization obtained several benefits, including saving money.

In [32] it is mentioned that inspections are a type of revision that is best suited to find design errors, while test techniques cannot find errors in the requirements documents.

On the other hand [33] mentions that the requirements validation techniques are important for defect identification. It also concludes and classifies that the inspection technique is an appropriate technique for large teams and organizations, which is expensive and where the client does not get involved.

3   Case Study Design

For this exploratory research, a case study (CS) was carried out considering the indications of [34] and [35]. Table 1 shows an adaptation of the CS phases at a high level.

Table 1. Phases and activities of the case study scheme proposed in [34]

Phase	Responsibility
P1	(P1-1) Objective of the Study
	(P1-2) What you Study
	(P1-3) Required Theory
	(P1-4) Research Questions
	(P1-5) Methods of Collection
	(P1-6) Data Selection
P2	(P2-1) Defining how Collection is Performed
E1	(E1-1) Data Collection
E2	(E2-1) Data Analysis
	(E2-2) Data Interpretation
R1	(R1-1) Reporting Results

Objective of the Study (P1-1). The objective of the study is to determine through a set of variables the benefit obtained from the validation of information system software requirements in the context of an information technology unit (ITU) in a public institution (government).

What is studied in this empirical research (P1-2) is the effect of process improvement, in particular, the adoption of software validation practices in the information systems development requirements phase in a set of six projects from an ITU of a public institution.

The public institution, whose identity is not revealed because of confidentiality agreements, is an entity dictating national policies and strategy on a main theme with benefits for the population, effectively articulating the state, civil society and the international community. The ITU has five specialists: an analyst, two programmers, a database administrator and a project manager. Depending on the priority and importance of the project, professionals can be incorporated to foster the project execution.

The analysis units of the study (see Table 2) correspond to a group of selected projects considering that all would begin their execution after the beginning of this investigation.

Table 2. List of selected projects

Project code	Software name
SI01	Software for the management of warehouse area processes
SI02	Software for the management of the purchasing process
SI03	Software for monitoring of the counseling service consultants
SI04	Software to prioritize projects financing applications and government activities
SI05	Software to manage the recruitment processes
SI06	Software for the supervision of physical and financial project goals

Required Theory (P1-3) is presented in Sect. 2.

Research Question (P1-4) established in this study is: Which is the degree of improvement (benefits) obtained as a result of the requirements validation process improvement? This question is broken down into the following study hypotheses:

- Requirements validation improves the schedule performance index (SPI) of information systems software development projects
- Requirements validation improves the cost performance index (CPI) of information systems software development projects
- Requirements validation improves the defect removal efficiency (DRE) in information systems software development projects.
- The validation of requirements increases the user satisfaction of the products of the information systems software development projects.

The schedule performance index (SPI) and cost performance index (CPI) are indicators to monitor and control the execution of projects, these are part of the earned value technique [36]. Defect removal efficiency (DRE) is a quality metric, proposed in [22], which provides benefits both at the project and process level. Satisfaction is defined as the value that users grant to a set of software properties. The properties evaluated are functional completeness, functional correction, functional appropriateness, ease of understanding, ease of learning, ease of operation, and satisfaction of use, these properties were extracted from the ISO/IEC 25000 quality model.

Method of Collection (P1-5) was carried out through three instruments developed for this purpose: (i) a software requirements document evaluation form (SRDEF) presented in Appendix A and whose formulas are detailed in Table 3; (ii) a project status report format (PSRF) presented in Appendix B and whose formulas are detailed in Table 4; and, (iii) a user satisfaction survey (USS) presented in Appendix C.

Table 3. Formulas defined in the SRDEF

Abbrev.	Index name	Formula
ISRS	Quality of the software requirements specification document	ISRS = a/b a: affirmative responses b: total questions
IPSR	Desirable properties of software requirements	IPSR = a/(b × c) a: sum of scores per requirement b: number of requirements c: maximum requirement score = 16
SRQI	Software requirements quality index	SRQI = ISRS * IPSR

The user satisfaction survey was developed based on the functional adequacy, usability and quality of use. These properties and characteristics are part of the ISO/IEC-25010 quality model proposed in [37].

Data Selection (P1-6). All the data obtained from the study is considered, which was based on a quasi-experimental design with posttest only. For this purpose, the SI01, SI03 and SI05 projects were considered as an experimental group, in which the new requirements validation procedure was applied (see Appendix D, improved process); and the SI02, SI04, SI06 projects as a control group, in which no changes were made to the way of working (see Appendix E, process without improvement). The design scheme is presented in Table 5.

The improved process for the validation of requirements (see Appendix D) considers four basic activities that are: planning, inspection, defect management and acceptance. In the planning phase the schedule was defined, the inspection instruments (check list and collecting form) were designed and the list of inspectors (users and technicians) was selected. In the next phase, the inspectors applied instruments to identify defects. The project manager prioritized the defects and the requirements

Table 4. Formulas defined in the PSRF

Abbrev.	Index name	Formula
SV	Schedule variation	EV-PV EV: Earned value PV: Planned value
SPI	Schedule performance index	EV/PV EV: Earned value PV: Planned value
CV	Variation in cost	EV-AC EV: Earned value AC: Actual cost
CPI	Cost performance index	EV/AC EV: Earned value AC: Actual cost
DRE	Defect removal efficiency	Eo/(Eo + Ei) Eo: Number of validation defects Ei: Number of accumulated defects

Table 5. Design scheme

Group	Treatment or stimulus	Results measurement
G-Exp.	Yes	O-Exp.
G-Con.	No	O-Con.

document was corrected and updated by analyst (defect management phase). A defect report and action performed were presented by analyst to inspectors (acceptance phase).

The stimulation or treatment of the experimental design is an improvement of the requirements validation traditional process. The improvement seeks to overcome the deficiencies in absence of strategy, non-standardized criteria for validation, absence of a formal validation technique, and lack of defects registration and control. The new requirements validation procedure considers tasks and instruments that support the activities of planning, inspection, defect management and acceptance.

The experimentation process had the following stages: selection and grouping of the software projects, application of the requirements validation improved process to the projects that are part of the experimental group, determination of the quality of the requirements specification document of all the projects in the sample, registry of defects and indicators of the earned value obtained during the execution of the project, and finally application of an user satisfaction survey in the production phase of the software.

4 Results

The values of ISRS, IPSR and SRQI were obtained from the application of the software requirements document evaluation form and detailed in Table 6. The first block shows the values of the experimental group and the second block shows the values of the control group. From the Table it can be seen that the projects in the experimental group (SI03, SI01, SI05) have higher rates than the projects in the control group (SI02, SI04, SI06). The indicators of the project execution as PV, EV, AC, SV, SPI, CV, CPI, Ei and DRE are presented in Table 7. The values detailed in the table correspond to the last monitoring and control observation of each project. This criterion was used because among the multiple observations, they represent better the final state of the project. The values of PV, EV, AC, SV and CV are expressed in Peruvian Soles thousands (kS/.).

Table 6. Software requirements quality index

Group	Project	ISRS	IPSR	SRQI
G-Exp.	SI01	0.83333	0.8819	0.7350
	SI03	0.91667	0.9267	0.8495
	SI05	0.75000	0.8940	0.6705
G-Con.	SI02	0.66667	0.7581	0.5054
	SI04	0.75000	0.6319	0.4740
	SI06	0.58333	0.6412	0.3740

With the values of Tables 6 and 7 three graphs were elaborated which explain the requirements quality influence on the cost, time, and quality of a project. Figure 1a shows that the CPI has a positive correlation with the SRQI (Pearson's r = 0.898 and p-value = 0.015), which shows that the cost performance does depend on the requirements quality. Figure 1b shows that the SPI has a positive correlation with the SRQI (Pearson's r = 0.976 and p-value = 0.001), which shows that the schedule performance depends on the requirements quality. Finally, Fig. 1c shows that the DRE has a positive correlation with the SRQI (Pearson's r = 0.978 and p-value = 0.001), which allows us to infer that the defects identification in early stages reduces the number of defects related to requirements in the stages of software construction and production rollout.

Table 8 shows the satisfaction expressed as a percentage of users that agree or totally agree with a set of characteristics of the software in production. To assess the characteristics, satisfaction surveys were used. From the tabulated data it is noticed that the SI06 project does not present results, this is explained because at the date of the application of the survey it was not in production. Table 8 shows N.A. (Not Available) for SI06, because it was not in production during the application of the satisfaction survey.

The results show that the average satisfaction for the experimental group is 75% while the average satisfaction for the control group is 36%. This significant difference

Table 7. Indicators of the project execution.

Group	Proj.	PV (kS/.)	EV (kS/.)	AC (kS/.)	SV (kS/.)	SPI	CV (kS/.)	CPI	Ei	DRE
G-Exp.	SI01	78.9	76.0	81.8	−2.8	0.963	−5.7	0.929	27	0.431
	SI03	73.5	72.3	78.8	−1.2	0.983	−6.4	0.917	31	0.425
	SI05	74.7	67.9	80.4	−6.8	0.908	−12.5	0.843	32	0.319
G-Con.	SI02	126.2	104.5	131.6	−21.7	0.828	−27.1	0.794	41	0.145
	SI04	206.3	158.9	277.3	−47.4	0.770	118.4	0.573	42	0.160
	SI06	98.3	67.4	156.6	−30.9	0.685	−89.1	0.430	32	0.085

(Student's T = 4.327 and p-value = 0.023) explains the requirements validation influence on the software products quality and how this is reflected in user satisfaction. From the results obtained it can be verified that requirements validation does achieve improvements in the defined variables, which implies that the hypotheses raised are true.

The validity of this study is presented in following lines:

Construct Validity. To guarantee the validity of the construct, quality of software requirements, the instruments that collect information to calculate the index that represents it were validated. The validation was carried out through expert judgment using criteria of clarity, objectivity, timeliness, organization, sufficiency, relevance, consistency, coherence and relevance. The measure used is the CVC content validity coefficient proposed by Hernández-Nieto, obtaining a coefficient of 0.8719 that qualifies the instruments and the construct as valid. On the other hand, constructs such as chronogram performance efficiency, cost performance efficiency are concepts widely studied in project management, for the study the PMI definitions were used. Similarly, the defect removal efficiency construct is a concept proposed by Pressman as an indicator of software product quality.

Internal Validity. There are variables that were not considered and that may affect the results. The efficiency in project planning, the managerial capacity of the project manager, the competencies and motivation of the technical team in charge and the support of senior management were not controlled. However, we must mention that the six projects in the sample were directed by different project managers, for all projects the members of the technical team went through a common learning curve and did not identify a differentiated participation of senior management in any of the projects.

External Validity. The study is limited to a single organization so it is not generalizable.

Reliability. The study was based on a quasi-experimental design, clearly identifying the independent variable, the stimulus or treatment and the dependent variables. The obtaining and analysis of results was supported by statistical methods with previously validated instruments that reduced or eliminated some bias of the evaluator.

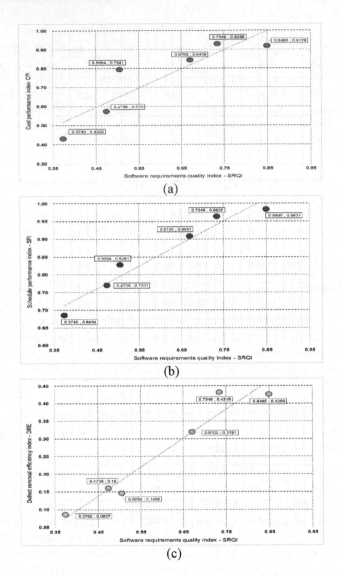

Fig. 1. Variable correlation graphs. (a) Relationship between the *software requirements quality index (SRQI)* and the *cost performance index (CPI)*. (b) Relationship between *software requirements quality index (SRQI)* and the *schedule performance index (SPI)*. (c) Relationship between the *software requirements quality index (SRQI)* and the *defect removal efficiency index (DRE)*.

Table 8. Percentage of users satisfied with information systems

Proj.	Functional complet.	Functional correct.	Functional approp.	Easy to underst.	Easy to learn	Easy to operate	Satisfac of use
SI01	70%	80%	30%	80%	90%	60%	70%
SI03	100%	78%	100%	33%	22%	78%	78%
SI05	88%	100%	75%	50%	100%	100%	100%
SI02	50%	50%	57%	50%	29%	29%	29%
SI04	40%	30%	20%	30%	60%	20%	10%
SI06	N.A.	N.A.	N.A.	N.A.	N.A.	N.A.	N.A.

5   Final Discussion and Future Work

In this work, it can be evidenced empirically that the introduction of software requirements validation has positive effects (benefits) in the projects execution, in particular, in terms of cost, schedule and quality. These results correspond to others in the context of information systems such as those presented in [31] and in [26]. This allows affirming that, obtaining requirements with quality, as the first technical product of the software development cycle, increases the possibility of software projects success.

The study adapted the inspection techniques to improve the requirements validation process and concordant with [27] the complexity of the projects, the experience of the development team and the size of the technical unit were considered.

The study shows that inspections, as a technique to validate software requirements, can be efficient in small teams and organizations, without generating additional costs and with the involvement of customers or users, contradicting those proposed in [33].

But it is a pending task to establish the efficiency of the techniques Pre-review and Requirements Reading mentioned in [26] as the most used validation techniques. It is proposed as future work to compare the efficiency and identify the conditions that allow optimizing their application.

Acknowledgments. This work is partially supported by Departamento de Ingeniería and the Grupo de Investigación y Desarrollo de Ingeniería de Software (GIDIS) from the Pontificia Universidad Católica del Perú.

Appendixes

https://drive.google.com/open?id=1f9NjU03ZHCJUodsgdZ3XVzih_yFkRsF2

References

1. PCM, RM.N. 041-2017-PCM. Uso obligado de la NTP-ISO/IEC 12207.2016, p. 2 (2017)
2. Fayad, M.E., Laitinen, M., Ward, R.P.: Software engineering in the small. Commun. ACM **43**(3), 115–118 (2000)

3. ISO/IEC: ISO/IEC TR 29110-1:2016 Systems and software engineering – Lifecycle profiles for Very Small Entities (VSEs) – Part 1: Overview, Geneva (2016)
4. Laporte, C., April, A., Renault, A.: Applying software engineering standards in small settings: historical perspectives and initial achievements. In: Proceedings of the First International Research Workshop for Process Improvement in Small Settings, pp. 39–51 (2006)
5. Paredes, L.M.: Validación del modelo liviano de gestón de requisitos para pequeñas empresas de desarrollo de software. GTI **9**, 73–84 (2010)
6. Oktaba, H., et al.: Modelo de Procesos para la Industria de Software, MoProSoft, Por Niveles de Capacidad de Procesos. Versión 1.3, Mexico DF, Mexico (2005)
7. SOFTEX: MPS. BR - Melhoria de Processo do Software Brasileiro. Guia Geral MPS de Software, Brasil (2012)
8. Hurtado, J., Bastarrica, C.: PROYECTO SIMEP-SW1. Trabajo de Investigación: Hacia una Línea de Procesos Ágiles Agile SPsL. Versión 1.0, Cauca, Colombia (2005)
9. CTN-ISSIGP: Normas Técnicos Peruanas. CTN-ISSIGP (2018). http://ctn-issi.pucp.pe/normas-tecnicas-peruanas
10. ISO/IEC: ISO/IEC TR 29110-5-1-2:2011 Software Engineering – Lifecycle Profiles for Very Small Entities (VSEs) – Part 5-1-2: Management and Engineering Guide: Generic Profile Group: Basic Profile, Geneva (2011)
11. Piattini, M., Calero, C., Moraga, M.A.: Calidad del Producto y Proceso Software, Madrid, España (2010)
12. Sommerville, I., Alfonso Galipienso, M.I., Botía Martinez, A., Mora Lizán, F., Trigueros Jover, J.P.: Ingeniería del Software, Septima. Madrid, España (2005)
13. Niazi, M., Mahmood, S., Alshayeb, M., Qureshi, A.M., Faisal, K., Cerpa, N.: Toward successful project management in global software development. Int. J. Proj. Manag. **34**(8), 1553–1567 (2016)
14. Hussain, A., Mkpojiogu, E.O.C., Kamal, F.M.: The role of requirements in the success or failure of software projects. Int. Rev. Manag. Mark. **6**(7Special Issue), 305–310 (2016)
15. Bhardwaj, M., Rana, A.: Key software metrics and its impact on each other for software development projects. Int. J. Electr. Comput. Eng. **6**(1), 242–248 (2016)
16. Toro, A.D.: Un Entorno Metodológico de Ingeniería de Requisitos para Sistemas de Información. Universidad de Sevilla (2000)
17. Tascon, C.A., Domínguez, H.F.: Análisis a la utilidad de la técnica de escenarios en la elicitación de requisitos. Rev. Antioqueña las Ciencias Comput. Ingeniería Softw. **7**, 59–67 (2017)
18. Derakhshanmanesh, M., Fox, J., Ebert, J.: Requirements-driven incremental adoption of variability management techniques and tools: an industrial experience report. Requir. Eng. **19**(4), 333–354 (2014)
19. Ivanov, V., Rogers, A., Succi, G., Yi, J., Zorin, V.: What do software engineers care about? Gaps between research and practice, pp. 890–895 (2017)
20. Rashid, J., Nisar, M.W.: How to improve a software quality assurance in software development-a survey. Int. J. Comput. Sci. Inf. Secur. **14**(8), 99–108 (2016)
21. ISO/IEC/IEEE: ISO/IEC/IEEE 24765:2017 Systems and software engineering – Vocabulary, Geneva (2017)
22. Pressman, R.S.: Ingeniería del Software. Séptima, México (2010)
23. ISO/IEC: ISO/IEC 12207:2008 Systems and Software Engineering – Software Life Cycle Processes, Geneva (2008)
24. De la Cruz-Londoño, C.A., Castro-Guevara, G.A.: Metodología Para la Adquisición y Gestión de Requerimientos en el Desarrollo de Software para Pequeñas y Medianas Empresas (Pymes) del Departamento de Risaralda. Universidad Tecnológica de Pereira (2015)

25. Maalem, S., Zarour, N.: Challenge of validation in requirements engineering. J. Innov. Digit. Ecosyst. **3**(1), 15–21 (2016)
26. Allasi,D., Dávila, A.: Financial impact on the adoption of software validation tasks in the analysis phase: a business case. In: Trends and Applications in Software Engineering, CIMPS 2017. Advances in Intelligent Systems and Computing, vol. 688, pp. 106–116 (2018)
27. Padilla Vedia, C.J.: El Desarrollo de Proyectos de Software y la Importancia de la Ingeniería de Requerimientos. In: Universidad Autónoma Juan Misael Saracho, vol. 2, pp. 14–24 (2017)
28. Toro Lazo, A.: Procedimiento para Especificar y Validar Requisitos de Software En Mipymes Desarrolladores de Software en la Ciudad de Pereira, Basado en Estudios Previos en la Región. Universidad Autónoma de Manizales (2017)
29. Zamuriano Sotés, R.F.: Las Inspecciones de Software y las Listas de Comprobación. Universidad del Valle Bolivia (2010)
30. Degiovanni, G.R.: Técnicas Automáticas para la Elaboración, Validación y Verificación de Requisitos de Software. Universidad Nacional de Córdova (2015)
31. Rodríguez Barajas, C.T.: Impacto de los requerimientos en la calidad de software. Tecnol. Investig. Acad. **5**(2), 161–173 (2017)
32. Alsayed, A.O., Bilgrami, A.L.: Improving software quality management: testing, review, inspection and walkthrough. Int. J. Latest Res. Sci. Technol. **6**, 1–12 (2017)
33. Bilal, H.A., Ilyas, M., Tariq, Q., Hummayun, M.: Requirements validation techniques: an empirical study. Int. J. Comput. Appl. **148**(14), 5–10 (2016)
34. Runeson, P., Höst, M.: Guidelines for conducting and reporting case study research in software engineering. Empir. Softw. Eng. **14**(2), 131–164 (2009)
35. Genero, M., Cruz-Lemus, J., Piattini, M.: Métodos de Investigación en Ingeniería de Software (2013)
36. PMI: Guía de los Fundamentos para la Dirección de Proyectos (Guía del PMBOK) Sexta Edición. EE-UU, Pensylvania (2017)
37. ISO/IEC: ISO/IEC 25010:2011 Systems and Software Engineering – Systems and Software Quality Requirements and Evaluation (SQuaRE) – System and Software Quality Models, Geneva (2011)

Towards a Social and Human Factor Classification Related to Productivity in Software Development Teams

Liliana Machuca-Villegas[1,2]($\boxtimes$) and Gloria Piedad Gasca-Hurtado[2]($\boxtimes$)

[1] Universidad del Valle, Calle 13 # 100-00, 760032 Cali,
Valle del Cauca, Colombia
liliana.machuca@correounivalle.edu.co
[2] Universidad de Medellín, Carrera 87 no. 30-65, 50026 Medellín, Colombia
gpgasca@udem.edu.co

Abstract. Software product development is characterized as an activity that focuses on social and human factors. In fact, studying these factors may be particularly appealing to software organizations seeking initiatives for fostering team productivity. In this light, the classification of the factors that may have an impact on the productivity of the software development team becomes the point of departure for the selection and definition of improvement strategies. As part of this research, we designed a methodology grounded in systematic literature review processes to approach a classification for social and human factors and develop a validation plan. In addition, the definition of inclusion and exclusion criteria yielded an approximation to the proposed classification, which may be used as the input for the definition of improvement actions. Finally, this work reports on the preliminary advances of the validation plan.

Keywords: Social and human factors · Software development productivity · Software development team · Literature review

1 Introduction

Software product development is characterized as a social activity [1–3] wherein social and human factors (SHF) are considered critical elements that have a clear impact on software project costs [4, 5]. In short, software product development is a process that demands teamwork skills, collaborative work skills, and high levels of motivation, which some authors refer to as "soft skills". These skills are supplemented by challenges such as group learning, team cohesion, autonomy that may affect the team performance and influence on the successful completion of software development projects [6, 7].

In this light, studying these factors may be particularly appealing to software organizations seeking initiatives for fostering productivity in their teams. For this purpose, this study proposes a social and human factor identification process based on a

© Springer Nature Switzerland AG 2020
J. Mejia et al. (Eds.): CIMPS 2019, AISC 1071, pp. 36–50, 2020.
https://doi.org/10.1007/978-3-030-33547-2_4

tertiary literature review [8]. Nevertheless, this process required the definition of a methodology based on systematic literature review processes [9]. The methodology led the SHF classification process. It is based on the selection of studies and data extraction from SLR protocol. However, additional tasks were proposed in order to identify and select of SHF. Finally, this methodology is used to implement four phases oriented towards the systematic and formal characterization of the SHF.

As part of its results, this research study investigated a classification for the SHF related to the productivity of the software development team. This classification may be used as a point of departure for the selection and definition of productivity-related strategies and improvement actions.

The rest of the paper is structured as follows. Section 2 lists the studies focusing on the SHF that may influence the productivity of software development teams. Section 3 describes the methodology used to characterize SHF. Section 4 presents the process execution results, leading to a preliminary classification of social and human factors. Section 5 discusses the results presented in the previous section. Section 6 presents the validation proposal. Section 7 mentions the threats to the classification process of SHF. Finally, Sect. 8 presents the conclusions drawn from this study.

2 Background and Related Work

This section provides a brief summary of some studies that focus on the classification of the factors that influence software development productivity and reports on their benefits.

Some of these studies on factors that influence software development productivity include systematic literature reviews as well as reviews of master's and PhD thesis papers. Therefore, for matters related to the search and classification of SHF, these studies have been used as the main input for this work [8]. However, as a supplementation strategy, this work also incorporates other studies that provide different dimensions to the studies previously selected from the original systematic literature reviews.

For example, some of the studies found include tertiary reviews on the factors that influence productivity [8]. As it was published in 2018, this study constitutes the most recent literature review on the subject. Based on the studies reviewed, this paper [8] asserts that no common classification of factors currently exists and alternatively propose categorizing factors as organizational and human factors.

However, Pirzadeh [2] presented the results of a systematic literature review (SLR) on the human factors in software development. The assessment of the main studies was aimed at identifying and characterizing the human factors that exert influence on the software development process from the standpoint of the software development cycle and management.

Likewise, Sampaio et al. [10] reviewed productivity factors and describe the strategies to maximize or minimize the influence of the factors, whether positive or negative. This classification is based on Boehm's proposal for the establishment of productivity boosting strategies in software product development.

In addition, a productivity study focused on developers at three different organizations was identified. However, rather than focusing on technical factors, this study describes other important factors, such as enthusiasm at work, peer support for new ideas, and receiving meaningful feedback, that affect performance [11].

In the context of agile software development, several factors exerting influence on productivity have also been observed. For example, Fatema and Sakib [12] use systems dynamics to reveal the factors affecting productivity in agile teams. Herein, the most outstanding factors discovered were team effectiveness, team management, motivation, and customer satisfaction.

Iqbal et al. [13] perform an empirical analysis on the factors that influence agile teams. This study describes the factors that arise at a certain frequency, such as inter-team relationships, team speed, team vision, and other factors related to the team member and lead roles.

However, even when these studies discuss the factors that influence software development productivity, they only focus on factor identification and generating recommendations for increasing team productivity. Consequently, these studies show a shortage of easy-to-implement techniques for addressing the factors that hinder software development productivity [1]. Further, to reduce production costs by increasing productivity, the organization must select and implement the practices based on its main influencing factors [14]; in turn, this implementation requires the use of appropriate techniques. Therefore, a classification of SHF particularly aimed at fostering the productivity of the software development team is a necessary input for establishing concrete and decisive improvement strategies and actions for SHF.

3 Methodology

This section proposes a methodology for the classification of SHF related to the productivity of the software development teams.

The objective of this methodology is to systematize and formalize the social and human factor categorization process based on previous classifications. Figure 1 denotes the methodology proposal in phases, tasks, cycles and iterations. The first two phases have been adopted from the SLR process proposed by Kitchenham [9]. The other phases come from cycles and iterations to facilitate feedback from the task results. This methodology is important because it led the SHF classification process. Through it, a systematic and iterative process has been followed to identify factors.

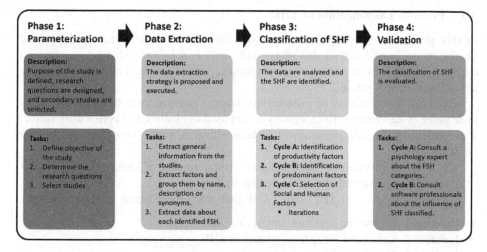

Fig. 1. Methodology for the classification of SHF

The different methodology phases are described below, along with the corresponding tasks, cycles, and iterations for each task.

3.1 Phase 1. Parameterization

This phase sets forth the starting parameters and establishes the need for studying the influence of SHF on the productivity of the software development team. In this phase, the following tasks are defined:

Task 1. Define objective of the study: establish the purpose of the study.

Task 2. Design research questions: as per the rules from Kitchenham's SLR protocol [9], the research questions required for proposing a research study must be designed.

Task 3. Select studies: define the study sample based on the parameters from the SLR protocol [9].

3.2 Phase 2. Data Extraction

This phase extracts data according to the parameters of interest defined from the research questions established in Phase 1. In this phase, the following tasks are defined:

Task 1. Extract general information from the studies: Based on the SLR protocol, general data, such as the names of the authors, country, keywords, factor classification adopted, number of factors identified, and abstract, must be extracted from the studies.

Task 2. Extract and group factors: assess the productivity factors reported by the selected studies in order to group them by name, description, or semantics.

Task 3. Extract data from SHF: analyze the SHF identified, extracting their name, description, and classification category.

3.3 Phase 3. Classification of SHF

In this phase, the SHF reported in the selected studies are identified and assessed. Owing to the importance of this phase and number of factors expected, we propose cycles and iterations to secure the corresponding classification. The following cycles are defined for this phase:

Cycle A. Identification of productivity factors: consolidation of factors based on their name or description. Here, synonyms are used to identify and consolidate factors. The goal is to identify productivity influence factors in the software development context.

Cycle B. Identification of predominant factors: number of factors from the classification categories identified in the selected studies.

Cycle C. Selection of SHF: select SHF and group them based on their semantic.

- Iteration 1. Define the selection criteria for SHF
- Iteration 2. Select the categories related to SHF
- Iteration 3. Group the factors by semantics or synonyms
- Iteration 4. Define the categories according to grouping.

3.4 Phase 4. Validation

This phase refines and evaluates the selection and classification of the SHF proposed from the methodology. This validation is performed in two cycles:

Cycle A. Consult an expert: categorize factors based on a semi-structured interview with psychology experts. The goal is to assess defined categories have an appropriate name and function according to characteristics of the SHF. It also seeks to assess that SHF belong to appropriate category.

Cycle B. Survey-based validation: define the influence of the SHF classified by associating them with the productivity of the software development team. This cycle is performed by software engineering experts and professionals.

4 Results

This section details the results obtained for each methodology phase.

4.1 Phase 1. Parameterization

This research study aims to identify the SHF that influence the productivity of the software development team. Based on this objective, three research questions were designed. Table 1 lists each question with its specific objective.

The study sample selected was defined based on the tertiary review from [8, 14]. This review focuses on the secondary studies that describe the factors that exert an influence on software productivity. In addition, these studies used a rigorous review protocol which facilitates their integration with other studies of the same nature. From this integration, a new analysis is proposed for the classification of factors into SHF.

Table 1. Research questions

ID	Question	Specific objective
RQ1	Which factors influence the productivity of the software team?	To identify the factors reported in the literature that exert an impact on the productivity of the software team
RQ2	Which are the predominant factors?	To identify leading factors based on their number of citations so that they may be used as predominant factors
RQ3	Which factors may be classified as SHF?	To identify social and human factor classification categories from the studies assessed and select potential factors for the proposal being developed

The four studies selected from the tertiary review [8] were part of this study sample. Furthermore, two other studies [2, 10], were added to this review owing to their relevance to this research study. Table 2 displays the studies selected with their corresponding identifiers.

Table 2. Secondary studies selected

ID	Title	Reference
ID-R1	A Systematic Review of Productivity Factors in Software Development	[15]
ID-R2	Factors Influencing Software Development Productivity - State-of-the-Art and Industrial Experiences	[16]
ID-R3	Factors that Influence the Productivity of Software Developers in a Developer View	[17]
ID-R4	What do we know about high performance teams in software engineering? Results from a systematic literature review	[18]
ID-R5	A Review of Productivity Factors and Strategies on Software Development	[10]
ID-R6	Human Factors in Software Development: A Systematic Literature Review	[2]

4.2 Phase 2. Data Extraction

The content analysis technique was used during the data extraction process [19]. This technique creates a coding scheme that translates important attributes or aspects extracted from the assessed studies into codes. For data systematization during this extraction, various Microsoft Excel forms were used.

Data were extracted from two types of assessments: general and specific level analyses. For both cases, a coding program was developed to facilitate the extraction of information.

Table 3 provides a sample template for extracting results at a general level. Table 4 provides an example of the analysis results at a specific level.

Table 3. Sample data extraction template at general level

ID	Title	Authors	Country	KW*	Years	Abstract	Main factors	Classification	Origin	Quantification*
R1	A systematic review of productivity factors in software development	Stefan Wagner & Melanie Ruhe	Germany	N/A	2008	Lists main factors from a literary review. Emphasizes the importance of assessing soft factors. Analysis from 1970 to 2007	Non-technical Factors Communication Developer skills	Technical Factors Non-technical Factors	Own classification	TF: (PF = 12, PrF = 8, DEF = 4) SF: (OCF = 3, TCF = 8, CEF = 8, WEF = 5, PjF = 3)

* KW: keywords, technical factors: TF, product factors: PF, process factors: PrF, development environment factors: DEF, soft factors: SF, organizational culture factors: OCF, team culture factors: TCF, capabilities and experience factors: CEF, work environment factors: WEF, project factors: PjF

Table 4. Sample data extraction template at specific level

Group	Category	Factor	CF*	Description
Social	Interpersonal	Communication (R1, R3, R4, R6)	4	"The degree and efficiency of which information flows in the team." (R1) "Appropriate communication in projects, frequent meetings to review the status, etc." (R3)
		Interpersonal Relationship (R3)	1	"Have a good relationship with colleagues and customers." (R3)
Personal	Capabilities and experiences	Analyst Capability (R1, R2)	2	"The skills of the system analyst." (R1)
		Applications Experience (R1, R2, R3)	4	"The familiarity with the application domain." (R1)"Knowledge on the domain of the application that is being developed. Example: Medical Systems." (R3)

* CF: Citation Frequency

4.3 Phase 3. Classification of SHF

This phase identifies the SHF from general factors that influence the software development productivity. This classification is based on the cycles defined in the methodology of Sect. 3. The corresponding results are presented below:

Cycle A. Identification of Generic Factors. The resulting classification distributes the factors into the proposed categories. As can be seen, some studies lean towards a technical approach [10, 15–17], while others lean towards a human approach [2, 18]. Table 5 provides a sample template generated with the factors identified.

Table 5. Sample generic factor identification template

Category		Factors
Technical factors	Product	Required Software Reliability (R1), Required reliability (R2)
	Process	Process Maturity (R1), Process maturity & stability (R2), Maturity Level (R3)
	Development environment	Programming Language (R1, R2, R3, R5)
Soft factors	Corporate culture	Credibility (R1)
	Team culture	Camaraderie (R1)
	Capabilities and experiences	Analyst Capability (R1, R2)
	Environment	Proper Workplace (R1), Work Environment (R3), Workstation (R3), organization environment (work environment) (R6)
	Project	Schedule (R1), Schedule pressure (R2), scheduling (R6)
	Individual characteristics	Commitment (R3, R6)
	Training	Training level (R2), Training (R3, R6), group training (R6), training classes and workshops (R6)
	Client participation	Customer/user involvement (R2), Client participation and experience (R5)
	Knowledge management	Knowledge Management (R3), Shared Information (R4), knowledge sharing (R6)

Cycle B. Identification of Predominant Factors. To identify the predominant factors, the factors were counted as per the classification categories from Cycle A. The corresponding results are presented in Table 6.

Table 6. Number of factors per category

Factor group	Category	Number of factors
Technical factors	Products	12
	Process	9
	Development environment	4
Subtotal technical factors		*25*
Non-technical factors	Corporate culture	7
	Team culture	25
	Capabilities and experiences	23
	Environment	6
	Project	13
	Individual characteristics	18
	Training	4
	Client participation	2
	Knowledge management	3
Subtotal non-technical factors		*101*
Total		*126*

Cycle C. Selection of SHF for Software Productivity. This cycle required the definition of iterations to facilitate factor selection and identify SHF. These iterations are described below:

Iteration 1. The definition of the selection criteria is described in Table 7. These criteria were classified as inclusion and exclusion.

Table 7. Selection criteria for SHF

ID	Criterion	Description
Inclusion criteria		
IC1	Dependence	− Focus on human characteristics − Steering towards the person or the software development team (social)
IC2	Group	− Keep factor that contains other factors grouped from Cycle A classification − Consolidate all factors related by similar terms into a single factor
IC3	Semantic	− Factor meaning and its relationship with social and human aspects
Exclusion criteria		
EC1	Factors not related to social and human aspects or human characteristics	
EC2	Factors without an explicit definition of source/origin	

Iteration 2. In this iteration, categories that group the factors identified in Cycle A under the IC1 inclusion criteria are assessed. As a result, the following categories are selected: Team culture, capabilities and experiences, and individual characteristics. Further, the knowledge management category is included as it describes the factors that could be related to cooperation within the work team.

Iteration 3. Based on the factors included in the selected categories, the first factor grouping process is performed to identify the factors that share social and human aspects. As a result, two groups of factors are defined: social group and personal group.

Iteration 4. Afterwards, a second grouping process is performed to identify factors with similar semantics and approaches and set forth a subcategorization within the two previously defined classification groups.

The results from iterations 3 and 4 are shown in Table 8.

5   Discussion

In order to achieve the proposed classification, techniques such as grouping by synonyms [15] or semantic similarity [18], were used. In addition, we used the results from the tertiary review [8] as a point of reference for grouping and classifying some factors (Table 8), and a classification structure was defined to facilitate the entire factor grouping and categorizing process. For these purposes, the structure proposed by [15], which categorizes the factors as technical and non-technical, was implemented. However, it became necessary to add more categories for those factors that did not fit into the existing ones. The new categories identified (individual characteristics, training, client participation, and knowledge management) were classified as non-technical factors, as per the objectives hereof. As part of this process, 126 factors were classified (Table 5).

Table 8. Preliminary categorization of SHF

	Category	Factor		Category	Factor
Social group	Interpersonal	Communication (R1, R3, R4, R6)	Personal group	People management	Commitment (R3, R6)
		Interpersonal Relationship (R3)			Change resistance (R6)
		Collaboration (R6)			Independence (R6)
		Cooperation (R6)		Emotional and Motivational	Motivation (R3, R4, R5, R6)
		Group work (R6)			Work satisfaction (R4)
		Group psychology (R6)		Cognitive	Intelligence (R4)
		Group learning (R6)			Learning skills and styles
		Self-efficacy (R4)			Mentality patterns (R6)
		Trust (R4)			Judgment (R6)
		Empathy (R4)			Bias (R6)
		Emotional Intelligence (R4)			Rationality (R6)
		Mutual respect (R4)			Abstract/reflective thinking (R6)
		Shared Information (R4)		Personality	Personality (R4), (R6)
		Knowledge sharing (R6)			Attitudes (R4)
	Team Culture	Camaraderie (R1)			Thinking styles (R6)
		Sense of Eliteness (R1)		Capabilities and experiences	Analyst Capability (R1, R2)
		Team Cohesion (R1, R2, R4)			Programmer Capability (R1, R2)
		Innovation (R1, R6)			Applications Experience (R1, R2, R3)
		Organizational Commitment (R4)			Platform Experience (R1, R2)
	Team Characteristics	Team Identity (R1)			Language and Tool Experience (R1, R2)
		Clear Goals (R1, R4)			Manager Capability (R1, R2)
		Team motivation and commitment (R2)			Manager Application Experience (R1)
		Autonomy (R4)			Experience (R2, R3, R6)
		Leadership Style (R4)			Knowledge (R4)
		Diversity (R1, R4)			Development skills (R6)
					Estimation skills (R6)
					Teamwork capabilities (R2)
					Integrator experience & skills (R2)
					Teamwork and communication skills (R2)
					Design experience (R2)
					Management quality & style (R2)

The predominant factors were identified by the number of factors (Table 6) included in each classification category. In cases in which a factor consolidated a group of factors owing to their similarity, only the consolidated factor was counted. Based on the classification defined and the number of factors by category, a greater number of non-technical factors may be observed, wherein the categories of team culture, capabilities and experiences, and individual characteristics stand out. The technical factor groups exhibit fewer categories than the non-technical factors, which may influence the number of factors deemed as technical (Table 6).

Moreover, this study assessed factors matching more than one factor during classification process. From this analysis, the following factors were identified as being the most prominent: programming language, communication, and motivation, which were assessed under the same name in four of the studies reviewed.

In addition, most of the SHF identified evidence a lower number of matches among the assessed secondary studies, while the technical factors exhibit a greater number of matches among them. The former indicates a tendency or interest towards technical factors.

The SHF used were selected on the basis of the defined iterations. During each iteration, the close observance of the definition for the SHF adopted herein became paramount: SHF may be defined as the characteristics that identify human beings based on their behavior from a social and individual perspective within the software development process [14]. From this definition, the selection criteria for the SHF as well as the factor groups by semantic or concept affinity were established.

In the first iteration (Cycle C, Phase 3), the categories classified as comprising non-technical factors were selected (Tables 5 and 6) since these categories meet the IC1 inclusion criteria. As a result, the following categories were selected: corporate culture, team culture, capabilities and experiences, environment, project, individual characteristics, training, client participation, knowledge management. Then, all technical factors for which we were unable to identify the social and human tendencies as per EC1 were discarded.

In the second iteration (Cycle C, Phase 3), all the categories selected were analyzed according to IC1. As a result, the following categories are selected: team culture, capabilities and experiences and individual characteristics, i.e., the categories that better represent human characteristics. These categories describe the factors from an individual and social standpoint (IC1). Further, the knowledge management category is included as it describes the factors that could be related to cooperation within the team: shared information and knowledge sharing.

However, the following categories are excluded as they are oriented towards corporate aspects, workplace environment, and projects without a direct dependence on the individual or the software development team (EC1): corporate culture, environment, project, training, and client participation.

Based on the factors included in the selected categories, the first factor grouping process is performed to identify the factors that share social and human aspects. These categories were defined in a third iteration based on the criteria listed in Table 7. Therefrom, two groups are obtained: factors with a social focus and factors with a personal or individual focus. In this iteration, the factors without an identified definition

(EC2) or those that did not comply with IC1 and EC1 were removed. Other factors were integrated through the application of IC2.

Finally, in iteration 4 and based on the classification from iteration 3, a second grouping process identified similar factors in terms of their semantics and approach. This process defined a new categorization within the two existing classification groups: (a) a social approach, grouping factors related to the interpersonal aspect, team culture, and team characteristics, and (b) an individual or personal approach, grouping cognitive, psychological, personality, capability, and experience. Table 8 denoted the social and human factor categories for their subsequent validation.

Part of the methodological proposal for a social and human factor classification related to the productivity of the software development team focuses on a validation process. As such, the following section further details the proposed validation plan.

6   Validation Plan for the Proposed Approach

To secure a final social and human factor classification, the set of factors identified must be validated. Therefore, as an empirical strategy, a survey-based approach will be implemented to receive feedback from academic and business level experts.

The validation proposal is distributed in cycles to purge and assess our current social and human factor classification, as follows:

1. Cycle A: Seek opinion from an expert in psychology on the defined factor categories.
2. Cycle B: Seek opinion from a group of software engineering professionals on the influence of the classified factors.

Through the instruments used in each cycle, opinions regarding the classification of the SHF in the different categories as well as on the influence of SHF on software development productivity from academic and business experts will be collected. Based on this information, we expect a final social and human factor classification and to be able to identify those factors that exert the greatest influence on productivity.

For Cycle A, the first meeting was held with a psychology professional who has been made aware of the project and results from the preliminary classification, as presented in Table 8. During this meeting, an overall review of factor classification and grouping categories was conducted. In addition, each factor was reviewed in order to confirm their affinity to the classification group: social and personal. Subsequently, the grouping categories were assessed according to their names and the factors contained therein.

From this preliminary review, we have discovered that some factors may belong to both the social and personal groups, as in the case of the empathy factor. Further, the team culture and team characteristics categories may be merged into the team culture category. Despite these preliminary results, we must continue working with the psychology expert to better streamline our classification of SHF.

As for Cycle B, progress has been made from the statistical point of view in the design of the survey and other consultation instruments.

7 Threats from the Preliminary Social and Human Classification Process

The preliminary identification process of SHF from the literature review may present the following threats to the validity of its results: researcher bias, study selection process issues, inaccurate data extraction, and factor classification issues.

To prevent researcher bias during the literature review process, the entire process was overseen and reviewed by another, more experienced researcher.

The review process used for social and human factor identification did not follow the complete SLR process. However, it was based on a tertiary review [8], which fully followed the SLR protocol in a rigorous manner. Although this review is not constituted as an SLR, it does include activities proposed by the SLR protocol to maintain the strict discipline followed in SLR processes. In addition, the proposed methodology supported the entire execution of this study.

To prevent data extraction errors or inaccuracies, we used the content analysis technique [19], which focuses on minimizing errors during data extraction.

Nonetheless, some bias may have been introduced during the factor classification process owing to the lack of details in the factor description extracted from the reviews, as well as to the identification of suitable factor grouping categories. To avoid this threat, we requested validation from a second researcher and reviewed the factor consolidation strategies used by other authors, such as factor consolidation by name, semantics, or description. In addition, we expect further validation from academic and business experts through survey-based research techniques.

8 Conclusions

This article presents a preliminary classification of the SHF that influence software development productivity. These factors have been identified from a sample of six secondary studies assessed through methodology established for the purposes of this research.

In addition, a social and human factor characterization methodology is defined in this paper. In fact, this study was conducted based on its phases, cycles, and tasks, thus evidencing the importance of the methodology implemented for the execution of the entire process.

Finally, the process identified a total of 57 SHF, which were classified into two groups: social and personal. For each group, a set of categories has been established for grouping the factors by semantics and function to facilitate their treatment for later use.

Based on the preliminary social and human factor classification as presented in this paper, there is a base set of factors that may be used to propose initiatives to increase the software development productivity. The results of this research will be used to propose strategies based on gamification to influence productivity.

The formalization of the validation plan of the SHF identified and its corresponding execution are part of the future work proposed within the framework of this research.

References

1. Yilmaz, M.: A software process engineering approach to understanding software productivity and team personality characteristics: an empirical investigation murat (2013)
2. Pirzadeh, L.: Human factors in software development: a systematic literature review. Master Science Thesis Compute Science Engineering, vol. 368 (2010)
3. Amrit, C., Daneva, M., Damian, D.: Human factors in software development: On its underlying theories and the value of learning from related disciplines. A guest editorial introduction to the special issue. Inf. Softw. Technol. **56**, 1537–1542 (2014). https://doi.org/10.1016/j.infsof.2014.07.006
4. Adolph, S., Hall, W., Kruchten, P.: Using grounded theory to study the experience of software development. Empir Softw Eng **16**, 487–513 (2011). https://doi.org/10.1007/s10664-010-9152-6
5. Fernández-sanz, L., Misra, S.: Influence of human factors in software quality and productivity. Ccsa **2011**, 257–269 (2011)
6. Standish Group: 2015 Chaos Report (2015)
7. Lombriser, P., Dalpiaz, F., Lucassen, G., Brinkkemper, S.: Gamified requirements engineering: model and experimentation. In: International Working Conference on Requirements Engineering: Foundation for Software Quality, pp 171–187. Springer, Cham (2016)
8. Oliveira, E., Conte, T., Cristo, M., Valentim, N.: Influence factors in software productivity - a tertiary literature review. In: Proceedings of 30th International Conference on Software Engineering and Knowledge Engineering, pp. 68–103 (2018). https://doi.org/10.18293/seke2018-149
9. Kitchenham, B., Charters, S.: Guidelines for performing systematic literature reviews in software engineering version 2.3. Engineering **45**, 1051 (2007). https://doi.org/10.1145/1134285.1134500
10. de Barros Sampaio, S.C., Barros, E.A., De Aquino, G.S., et al.: A review of productivity factors and strategies on software development. In: Proceedings of 5th Intenational Conference on Software Engineering Advances, ICSEA 2010, pp. 196–204 (2010). https://doi.org/10.1109/icsea.2010.37
11. Murphy-hill, E., Jaspan, C., Sadowski, C., et al.: What predicts software developers' productivity ? IEEE Trans. Softw. Eng. 1–13 (2019, early access)
12. Fatema, I., Sakib, K.: Factors influencing productivity of agile software development teamwork: a qualitative system dynamics approach. In: Proceedings of Asia-Pacific Software Engineering Conference, APSEC 2017, pp. 737–742 (2018). https://doi.org/10.1109/apsec.2017.95
13. Iqbal, J., Omar, M., Yasin, A.: An empirical analysis of the effect of agile teams on software productivity. In: 2nd International Conference on Computing, Mathematics and Engineering Technologies (iCoMET), pp 1–8. IEEE (2019)
14. Cunha De Oliveira, E.C.: Fatores de influência na produtividade dos desenvolvedores de organizaciones de software (2017). https://tede.ufam.edu.br/bitstream/tede/6137/5/Tese_Edson C. C. Oliveira.pdf
15. Wagner, S., Ruhe, M.: A systematic review of productivity factors in software development. In: Proceedings of the 2nd International Software Productivity Analysis and Cost Estimation (SPACE 2008), pp. 1–6 (2008)
16. Trendowicz, A., Münch, J.: Factors influencing software development productivity-state-of-the-art and industrial experiences. Adv. Comput. **77**, 185–241 (2009). https://doi.org/10.1016/s0065-2458(09)01206-6

17. Paiva, E., Barbosa, D., Lima, R, Albuquerque, A.: Factors that influence the productivity of software developers in a developer view. In: Innovations in computing Sciences and Software Engineering (2010). https://doi.org/10.1007/978-90-481-9112-3
18. Dutra, A.C.S., Prikladnicki, R., França, C.: What do we know about high performance teams in software engineering? Results from a systematic literature review. In: 41st Euromicro Conference on Software Engineering and Advanced Applications (2015)
19. Bryman, A.: Social Research Methods. Oxford University Press, Oxford (2012)

Software Product Quality in DevOps Contexts: A Systematic Literature Review

Daniel Céspedes[1], Paula Angeleri[2], Karin Melendez[3],
and Abraham Dávila[3(✉)]

[1] Escuela de Posgrado, Pontificia Universidad Católica del Perú, Lima, Peru
dhcespedes@pucp.edu.pe
[2] Facultad de Ingeniería y Tecnología Informática, Universidad de Belgrano,
Buenos Aires, Argentina
paula.angeleri@comunidad.ub.edu.ar
[3] Departamento de Ingeniería, Pontificia Universidad Católica del Perú,
Lima, Peru
{kmelendez,abraham.davila}@pucp.edu.pe

Abstract. DevOps is a change in the organizational culture that aims to reduce the gap between development and operation teams, accelerating the software release process. However, little is known about the impact of this approach on software product quality. This study aims to analyze the influence of the application of DevOps on software product quality; therefore, a systematic literature review was conducted. Thirty-one articles related to DevOps and its influence on product quality were identified. The studies indicate a strong influence of some product quality characteristics, specifically: Reliability and Maintainability. Additionally, practices associated with DevOps, such as the minimum viable product, deployment automation, test automation, cloud computing and team cooperation, show a relationship with the improvement in software product quality, however, its adoption also brings new challenges to preserve security.

Keywords: Systematic literature review · DevOps · Product quality · ISO/IEC 25000

1 Introduction

Over the years, the software industry has had a significant change, from extensive projects that could last for years to an approach of short and continuous cycles [1]. According to [2], the execution of these short cycles, applying certain practices to prevent defects, can improve the product quality. However, [3] mention that their systematic mapping study do not show a relevant difference in the software product quality when using these practices or not. It should be noted that none of these studies mention standards or models of product quality.

On the other hand, this new approach, which promotes dynamic and continuous development and deployment practices, affects the daily work of the operations team, who require a stable environment [4]. Due to these differences in the development and

© Springer Nature Switzerland AG 2020
J. Mejia et al. (Eds.): CIMPS 2019, AISC 1071, pp. 51–64, 2020.
https://doi.org/10.1007/978-3-030-33547-2_5

operations teams, the DevOps paradigm emerges as a solution proposal [5]. DevOps, by the acronym of "Development and Operations", aims to change the organizational culture so the teams involved can align themselves to the same goal [5], offering the highest possible value to their organization, in a context of continuous changes and without impacting business continuity [6].

Although there are publications related to DevOps in the industry (empirical contexts), its impact on the software product quality has not been studied in detail. Therefore, the objective of this work is to identify the influence of DevOps on product quality through a systematic literature review (SLR). Also, we aligned this SLR with the model proposed in the ISO/IEC 25010 standard, in which quality model is a decomposition of characteristics and sub characteristics [7]. The article is organized in 5 sections: in Sect. 2, presents background and related work; in Sect. 3, describes the research method; in Sect. 4 presents the analysis of results; in Sect. 5, shows the final discussion and future work.

2 Background and Related Work

DevOps is a paradigm that promotes a relationship between the development and operation teams, working together to offer the greatest possible value to the organization, in a context of continuous changes on software product [5]. The use of the term DevOps began in 2008 when Patrick Debois, at the Agile 2008 Conference, mentioned the need for an agile infrastructure and interaction between the development and operations teams [8]. In 2009, the term DevOps became popular with the beginning of events called DevOpsDays [9].

The DevOps paradigm focuses mainly on making cultural changes in the organization, relying on approaches such as Continuous Delivery, which is aimed at the automation and optimization of the delivery process [6]. The biggest difference between DevOps and Continuous Delivery is that the first one focuses on the synergy between the development and operations areas, seeking the optimization of the product, as well as the continuous improvement of it [6].

In addition, there are different frameworks, processes and methodologies designed for DevOps application, some focused on automation: [10, 11] and others on security and risk management: [12, 13], as example of the context variety. However, in a preliminary review of the literature, few studies were identified that show results of empirical cases of DevOps application as [14, 15].

Furthermore, during 2005 the first part of the ISO/IEC 25000 standard series was published, as a result of integrating the best practices and lessons learned from previous standards, mainly ISO/IEC 9126 and ISO/IEC 14598 [16]. This standard series, specifically division 2501n, defines the software product quality model [7], which is the case of interest to this SLR. The software quality characteristics defined in the model are: Functional suitability, Performance efficiency, Compatibility, Usability, Reliability, Security, Maintainability and Portability.

Consequently, the aim of this SLR is to know the relationship between DevOps and the quality defined according to the software product quality model of ISO/IEC 25010. For that purpose, Table 1 presents the main studies related to the objective of this SLR.

From the above, no work was found that focuses on the DevOps relationship with the characteristics of some quality models and empirical case studies.

Table 1. Related works with the SLR

Title	Brief description
Adopting DevOps practices in quality assurance [17]	Describe the context and antecedents that originate the DevOps paradigm and how this change in the organizational culture impacts the development and operation teams
Continuous deployment of software intensive products and services: a systematic mapping study [3]	Systematic mapping that aims to analyze and classify studies related to continuous deployments, identifying the state of the art, reported results, areas for future research, application use benefits and related challenges
Report: DevOps literature review [18]	Systematic mapping that aims to identify the characteristics, related effects and factors that support DevOps, as well as detailing relevant concepts related to it
State of DevOps report (2014–2018) [19–23]	Annual reports that present the situation of DevOps in companies, according to a qualitative study through surveys on people related to this approach

3   Research Method

In this section, the protocol used for the SLR is shown. Following the methodology suggested in [24, 25] this review aims to answer the proposed research questions, based on publications stored in indexed digital databases.

3.1   Research Purpose and Questions

To recognize the purpose of the research, as shown Table 2, one template from Goal-Question-Metrics approach was used to define the objectives [26]. Thus, Table 3 presents the research and motivation questions for the SLR.

3.2   Review Protocol

Search Strategy. The PICO method was used for the definition of the search string [27] as shown Table 4. The intervention concept includes the terms "Development and Operations" and "DevOps" (abbreviation). In addition, the outcome concept contains the terms: "test", "quality", "verification", "validation" and quality characteristics of ISO/IEC 25010 standard. The following indexed databases were selected due to their relevance in Software Engineering [28]: ACM, IEEE Xplore, Scopus, and Science Direct. On the other hand, the search was limited since 2008, when the term DevOps began to be used [18]. The search was conducted in February 2019 and the search string for each database is shown in Appendix A.

Selection Criteria. The inclusion and exclusion criteria are presented in Table 5. In addition, three stages were implemented to refine the list of selected studies. The criteria used in each stage are shown in Table 6.

Quality Assessment. Following [24], the checklist for the quality assessment of studies in Table 7 was defined. The checklist for qualitative studies from [24] was used as a reference since the review focuses on empirical cases. The qualification used to assess compliance with each criterion was: Yes = 1, Partly = 0.5 or No = 0.

Data Extraction. For the extraction of the data, a template was used to answer the research questions in each selected study. The results obtained were tabulated in this template [24], facilitating the synthesis of the information and the answer to the research questions.

Table 2. Research purposes

Field	Value
Object of study	DevOps application
Purpose	Identify its influence
Focus	Software product quality
Stakeholder	Companies, governments, academia and professionals
Context	Software engineering projects

Table 3. Research questions

Research questions	Motivation
RQ1. Which product quality characteristics are reported in studies related to DevOps?	Identify the software product quality characteristics related to DevOps
RQ2. Which DevOps practices have influence on the software product quality?	Identify whether the application of DevOps practices helps or not in the software product quality
RQ3. Which considerations must be taken to improve product quality when adopting DevOps?	Identify the considerations to keep in mind when adopting DevOps in relation to software product quality improvements

Table 4. Search terms

Concept	Related terms
Population	"software"
Intervention	"devops" OR "development and operations"
Comparison	Not applied because it is not necessary to make a comparison with a pattern or reference
Outcome	"test*" OR "quality*" OR "functiona*" OR "suitab*" OR "perform*" OR "effic*" OR "compatib*" OR "usab*" OR "reliab*" OR "secur*" OR "maintainab*" OR "portab*" OR "verif*" OR "valid*"

Data Synthesis. We used two strategies for data synthesis. In the first case, we applied content analysis and the sub-characteristics level was used as a structure of categories on which to organize the selected studies. In the second case, we applied narrative synthesis.

Table 5. Inclusion and exclusion criteria

Item	Criteria	Type
CI.1	Studies registered in indexed databases	Inclusion
CI.2	Studies related to Software Engineering	Inclusion
CI.3	Studies that describe a relationship between DevOps and software product quality	Inclusion
CI.4	Studies that show empirical cases study	Inclusion
CE.1	Duplicated items	Exclusion
CE.2	Studies in a language other than English	Exclusion
CE.3	Secondary studies	Exclusion
CE.4	Studies that are not related to DevOps	Exclusion
CE.5	Studies that are not related to software product quality	Exclusion

Table 6. Stages and selection criteria

Stages	Scope	Selection criteria
First stage	Title	CI.1, CI.2, CI.4, CE.1, CE.2, CE.3
Second stage	Abstract and conclusions	CI.2, CI.4, CE.2, CE.3, CE.4, CE.5
Third stage	Complete content	CI.3, CI.4, CE.4, CE.5

Table 7. Quality evaluation criteria

Item	Evaluation criteria
EC.1	How credible are the findings?
EC.2	If credible, are they important?
EC.3	How has knowledge or understanding been extended by the research?
EC.4	How well does the evaluation address its original aims and purpose?
EC.5	How well is the scope for drawing wider inference explained?
EC.6	How well was data collection carried out?
EC.7	How well has the approach to, and formulation of, analysis been conveyed?
EC.8	How well are the contexts and data sources retained and portrayed?
EC.9	How clear are the links between data, interpretation and conclusions?
EC.10	How clear and coherent is the reporting?
EC.11	How clear are the assumptions/theoretical perspectives/values that have shaped the form and output of the evaluation?

4 Analysis of Results

In this section the results obtained from SLR are shown. Following the defined research method, 2195 studies were obtained using the search string in the indexed databases. Then, the selection process was executed in three stages, following their inclusion and exclusion criteria. Table 8 shows how the studies were processed and Table 9 lists the selected documents.

Table 8. Analysis by selection criteria

Criteria	First stage	Second stage	Third stage
Base	2195	621	437
CI.1	–	–	–
CI.2	–786	–34	–
CI.3	–	–	–91
CI.4	–	–2	–315
CE.1	–719	–	–
CE.2	–	–	–
CE.3	–14	–5	–
CE.4	–55	–132	–
CE.5	–	–11	–
Selected	621	437	31

Afterwards, the quality of the selected studies was evaluated, having as a result that all exceeded 50% of the maximum score. Finally, the information was extracted using the template defined in the previous section to answer the research questions.

4.1 RQ1. Which Product Quality Characteristics Are Reported in Studies Related to DevOps?

According to selected studies, the Table 10 presents the identified quality characteristics and sub characteristics mapped to studies selected. At the first line, labeled "Charact", characteristics considered were: reliability, maintainability, functional suitability, security, and performance efficiency (indicated in the Table as PE*). At the second line, labeled "Sub Charact", sub characteristics reported in the studies selected were considered. At the third line, labeled "Qty", show quantity studies in every sub characteristics. The rest of the Table shows the maps between studies and sub characteristics. From the review it was found that reliability (availability) and maintainability (testability and modifiability) are the most referenced characteristics related to DevOps.

4.2 RQ2. Which DevOps Practices Have Influence on the Software Product Quality?

Deployment Automation. The studies reviewed show a strong relationship between deployment automation with reliability and maintainability. Regarding reliability, S01, S03, S05, S06, S07, S08, S12, S13, S15, S16, S17, S19, S21, S22, S23, S26, S28, S29, S30 and S31 all show that DevOps improves the availability and recoverability by reducing the time of production deployments.

Table 9. Selected studies

Item	Study title
S01	Development and Deployment at Facebook [15]
S02	DevOps Patterns to Scale Web Applications using Cloud Services [29]
S03	ResearchOps: The Case for DevOps in Scientific Applications [30]
S04	Achieving Cloud Scalability with Microservices and DevOps in the Connected Car Domain [31]
S05	Chaos Engineering [14]
S06	DevOps Advantages for Testing: Increasing Quality through Continuous Delivery [32]
S07	Management Challenges for DevOps Adoption within UK SMEs [33]
S08	On the Impact of Mixing Responsibilities Between Devs and Ops [34]
S09	Towards DevOps in the Embedded Systems Domain: Why is It so Hard? [35]
S10	DevOps: Reliability, Monitoring and Management with Service Asset and Configuration Management [36]
S11	From Agile Development to DevOps: Going Towards Faster Releases at High Quality - Experiences from an Industrial Context [37]
S12	DevOps: Fueling the evolution [38]
S13	Quality Assessment in DevOps: Automated Analysis of a Tax Fraud Detection System [39]
S14	Towards Continuous Delivery by Reducing the Feature Freeze Period: A Case Study [40]
S15	Towards DevOps in Multi-provider Projects [41]
S16	Adopting Autonomic Computing Capabilities in Existing Large-Scale Systems: An Industrial Experience Report [42]
S17	Breaking Down the Barriers for Moving an Enterprise to Cloud [43]
S18	Building a Virtually Air-gapped Secure Environment in AWS* with Principles of DevOps Security Program and Secure Software Delivery [44]
S19	Building Science Gateway Infrastructure in the Middle of the Pacific and Beyond: Experiences using the Agave Deployer and Agave Platform to Build Science Gateways [45]
S20	Characteristics of Defective Infrastructure as Code Scripts in DevOps [46]
S21	Continuous Integration and Delivery for HPC: Using Singularity and Jenkins [47]
S22	DevOps Capabilities, Practices, and Challenges: Insights from a Case Study [48]
S23	Exploiting DevOps Practices for Dependable and Secure Continuous Delivery Pipelines [49]
S24	Identity and Access Control for micro-services based 5G NFV platforms [50]
S25	Implementation of a DevOps Pipeline for Serverless Applications [51]
S26	Implementing DevOps practices at the control and data acquisition system of an experimental fusion device [52]

(*continued*)

Table 9. (*continued*)

Item	Study title
S27	Making Runtime Data Useful for Incident Diagnosis: An experience report [53]
S28	Managing a Cray supercomputer as a git branch [54]
S29	Teaming with Silicon Valley to Enable Multi-Domain Command and Control [55]
S30	Implementing DevOps in Legacy Systems [56]
S31	Continuous Deployment at Facebook and OANDA [57]

Table 10. Software product quality attributes and related studies.

Charact. / Sub Charact.	Reliability			Maintainability					Functional Suitability		Security				P. E*
	Maturity	Availability	Recoverability	Modularity	Reusability	Analyzability	Modifiability	Testability	Functional Completeness	Functional Correctness	Confidentiality	Integrity	Accountability	Authenticity	Resource utilization
Qty	1	23	1	2	4	7	15	18	2	10	4	3	1	2	3
S01		X	X	X			X	X							
S02		X					X								X
S03		X			X		X	X							X
S04				X	X		X	X		X					
S05		X				X	X	X		X					
S06		X					X	X		X					
S07		X					X	X							
S08		X						X							
S09										X					
S10		X				X	X								
S11										X					
S12		X				X	X								
S13		X					X	X		X					
S14		X						X							
S15		X						X							
S16		X				X	X								
S17		X					X				X				
S18											X	X	X	X	
S19		X			X		X	X							
S20						X	X				X	X			
S21							X								X
S22		X							X	X					
S23	X	X					X	X							
S24													X	X	
S25										X					
S26		X							X	X					
S27		X				X	X	X							
S28		X					X	X							
S29		X					X	X			X				
S30		X				X	X			X					
S31		X			X										

Furthermore, in relation to maintainability, S01, S03, S04, S05, S06, S07, S12, S15, S16, S19, S21 and S29 explain that the automated deployments facilitate the attention of incidents and, together with the automated tests, improve the analyzability, as well as their modifiability and testability.

In addition, it should be noted that S17, S18, S20 and S29 state the importance of security, because it may be compromised when attempting to automate the deployment process.

Test Automation. As stated above, test automation is related to the continuous delivery process. Additionally, a relationship with reliability, maintainability and safety was found.

In the case of reliability, S05, S09, S13, S14, S15, S17, S19, S21, S22, S26, S27, S30 and S31 state that by facilitating the execution of the tests, the recoverability and availability are improved. In addition, by having a set of tests that is constantly improved, the maturity of the system also improves.

Considering the maintainability, according to S01, S05, S06, S13, S14, S15, S19, S29 and S30; the list of automated tests reduces the complexity of their preparation, improving their ability to be tested with each deployment. Furthermore, if this list considers safety tests, it results in an improvement of the various attributes of this quality characteristic, as mentioned in S18 and S29.

Personal Responsibility and Team Cooperation. The studies mention different proposals to reduce the gap between the operations and development team. The main strategies identified were: (1) Personal responsibility: The person or team takes responsibility for the entire process, from development to operation, (2) Team cooperation: Participation of those responsible for development and operations in the same team.

Given these points, S01, S02, S07, S08 and S09 mention that these collaboration strategies had an influence on reliability by having deployed changes in less time. In the same way, S01, S04, S13 and S17 indicate an influence on maintainability by improving modifiability and testability.

In the case of personal responsibility, S01, S02, S08 and S09 explain the positive effect on maintainability by improving their modularity, since they avoid dependencies between functionalities, reducing the impact between them. However, in S08 and S31 also state the negative impact on reusability by the duplication of functionalities due to the lack of coordination between different teams.

Cloud Computing. It is related to DevOps because it reduces the gap between the development and operations teams by simplifying the infrastructure operational complexity, reaching the point of being treated as one more development task (Infrastructure as Code). With regard to cloud computing, S08, S17, S19, S21, S22 and S30 indicate a relation with performance efficiency, since it improves the resources utilization by the intrinsic flexibility that the platform offers. Moreover, its modifiability is enhanced because is scalable on demand, besides its reusable by offering configuration scripts which quickly enables new instances.

Monitoring and Feedback. According to S05, S10, S16, S22, S25, S27 and S30, a positive influence on maintainability is observed by facilitating the identification and determination of the issues cause, having as a result an improvement of their analyzability. In addition, it allows improving the communication between the development and operations teams by establishing detection and attention mechanisms, identifying their common needs.

4.3   RQ3. Which Considerations Must Be Taken to Improve Product Quality when Adopting DevOps?

Continuous Deployment as a Starting Point for the Adoption of DevOps. Although it is mentioned that there is no recipe for the adoption of DevOps, S01, S03, S04, S05, S06, S07, S08, S12, S13, S15, S17, S18, S19, S21, S23, S26 and S29 point out as a common factor, the implementation of a platform for continuous deployments. This starting point could offer better results in product quality if continuous tests are used. Additionally, S25 and S30 acknowledge continuous deployment as a relevant factor in the DevOps adoption.

Additional Care in Security. The use of Cloud computing brings new challenges to preserve security in the system. According to S17 and S18, some proposals were presented to mitigate the risk: (1) Keep the sensitive information of the system on a separate server, (2) Limit administrative access on production environments to some managers, (3) Use a semi-automatic pipeline that prevents production deployments unless they have the necessary approvals, (4) Prepare a set of automated tests that prevent potential security risks, and (5) Review the configuration scripts to avoid compromising confidential information and any other security breach in the environment.

Chaos Engineering. Mentioned in S05, as a testing strategy. In this case consists of design test cases according to events or issues that could occur in a production environment. For this purpose, four principles are proposed: (1) Build a hypothesis about the behavior of the stable state, (2) Alter the events of the real world, (3) Execute the experiments in production, and (4) Automate the experiments for their continuous execution.

Not always Automated Testing. Not all tests are convenient to automate; therefore, strategies must be proposed to deal with semiautomatic or manual tests, without obstructing the flow of continuous deployment. S01, S05 and S06 specified, automated or not, the following tests and metrics: Chaos engineering, A/B testing, Code coverage, Mutation testing, Static analysis, Negative testing, Exploratory testing and Parallel testing.

Minimum Viable Product. S13, S19 and S29 recommend an initial release of a basic product, but sufficiently functional to offer a preliminary review. This practice allows a quick feedback from the user, as it facilitates the product quality tests preparation.

4.4   Threat of the Validity

The scheme worked for this SLR included the support of two researchers: an advisor and a collaborator. The following steps were performed: (i) The planning stage was executed with the guidance of the research advisor and the final planning document was reviewed by both researchers (co-authors); then, the comments raised were resolved; (ii) In the studies selection stage, some databases were left out because their search mechanisms are cumbersome and the validation of the selected studies was performed randomly by the research collaborator; (iii) The synthesis of the work was reviewed by the advisor research. In step (i) several approaches were made because there were no studies related with DevOps and quality models specifically, but there was a relationship between DevOps and certain quality characteristics. Therefore, we chose to work with the quality model of ISO/IEC 25010 and not with ISO/IEC 9126.

5   Final Discussion and Future Work

The purpose of this SLR was to identify the influence of the use of DevOps on product quality, using the ISO/IEC 25010 model as a reference. The findings show that there is an influence on product quality by the use of DevOps, mainly related on reliability and maintainability; and to a lesser extent functional suitability, security and performance efficiency. From the above, it is emphasized that the adoption of DevOps could add new considerations in security assurance.

In addition, the main practices associated with DevOps that influence the product quality are: deployment automation, test automation, cloud computing and team cooperation. However, Deployment automation and Cloud computing should be taken a greater care with security. Similarly, about team cooperation, communication inside and outside the team is relevant to avoid inconveniences with maintainability.

On the other hand, to begin with the adoption of DevOps impacting product quality, it was observed that the implementation of a continuous deployment process can offer relevant results regarding product quality. In the same way, the use of the development of a minimum viable product can offer an agile feedback, facilitating the development of tests cases. In addition, using tests, automatic or not, integrated into the deployment automation improve significantly the product quality.

Finally, according to what is analyzed in the SLR, it is recommended as future work to perform a systematic mapping to recognize the best practices in the adoption of DevOps on study cases. In addition, due to the few studies identified, it is recommended to perform more study cases, preferably using a quantitative analysis, in a business environment, with the aim of having a greater base of reference for the analysis and, therefore, improve the proposals for the adoption of this paradigm.

Acknowledgments. This work is framed within the CalProdSw project supported by the Department of Engineering and the Grupo de Investigación y Desarrollo de Ingeniería de Software (GIDIS) from the Pontificia Universidad Católica del Perú.

Appendix

https://drive.google.com/open?id=11yehxYJJ0VfJrgkbe0k2T_pYIrq0gZbp

References

1. Järvinen, J., Huomo, T., Mikkonen, T., Tyrväinen, P.: From agile software development to mercury business. In: LNBIP, vol. 182 (2014)
2. Chen, L.: Continuous delivery: huge benefits, but challenges too. IEEE Softw. 32(2), 50–54 (2015)
3. Rodríguez, P., et al.: Continuous deployment of software intensive products and services: a systematic mapping study. J. Syst. Softw. 123, 263–291 (2017)
4. Callanan, M., Spillane, A.: DevOps: making it easy to do the right thing. IEEE Softw. 33(3), 53–59 (2016)
5. Ebert, C., Gallardo, G., Hernantes, J., Serrano, N.: DevOps. IEEE Softw. 33(3), 94–100 (2016)
6. Virmani, M.: Understanding DevOps & bridging the gap from continuous integration to continuous delivery. In: 5th International Conference on Innovative Computing Technology, INTECH 2015, pp. 78–82 (2015)
7. ISO: International Standard ISO/IEC 25010 - Systems and software Quality Requirements and Evaluation (SQuaRE) - System and software quality models. ISO, Geneva (2011)
8. Debois, P.: Agile Infrastructure and Operations (2008)
9. Rapaport, R.: A short history of DevOps (2014). http://www.ca.com/us/rewrite/articles/devops/a-short-history-of-devops.html. Accessed 28 Dec 2017
10. Weerasiri, D., Barukh, M.C., Benatallah, B., Cao, J.: A model-driven framework for interoperable cloud resources management. In: Lecture Notes Compute Science (Including Subseries Lecture Notes Artificial Intelligence on Lecture Notes Bioinformatics). LNCS, vol. 9936, pp. 186–201 (2016)
11. Yoshii, J., Hanawa, D.: Technology to automate system operations for flexibly responding to dynamically changing cloud environments. Fujitsu Sci. Tech. J. 51(2), 81–85 (2015)
12. Díaz, O., Muñoz, M.: Implementation of a DevSecOps + Risk management in a data center of a mexican organization [Implementación de un enfoque DevSecOps + Risk Management en un Centro de Datos de una organización Mexicana]. RISTI – Rev. Iber. Sist. e Tecnol. Inf. 26, 43–53 (2018)
13. Díaz, O., Muñoz, M.: Reinforcing DevOps approach with security and risk management: an experience of implementing it in a data center of a mexican organization. In: 2017 6th International Conference on Software Process Improvement (CIMPS), pp. 1–7 (2017)
14. Basiri, A., et al.: Chaos engineering. IEEE Softw. 33(3), 35–41 (2016)
15. Feitelson, D., Frachtenberg, E., Beck, K.: Development and deployment at Facebook. IEEE Internet Comput. 17(4), 8–17 (2013)
16. ISO/IEC: ISO/IEC 25000: 2014 - Systems and software engineering – Systems and software Quality Requirements and Evaluation (SQuaRE) – Guide to SQuaRE (2014)
17. Roche, J.: Adopting DevOps practices in quality assurance. Commun. ACM 56(11), 38–43 (2013)
18. Erich, F., Amrit, C., Daneva, M.: Report: DevOps Literature Review, October 2014
19. Velasquez, N.F., Kim, G., Kersten, N., Humble, J.: 2014 State of DevOps report. Puppetlabs (2014)
20. Forsgren, N., Kim, G., Kersten, N., Humble, J.: 2015 State of DevOps report. Puppetlabs (2015)

21. Brown, A., Forsgren, N., Humble, J., Kersten, N., Kim, G.: 2016 state of DevOps. Puppetlabs (2016)
22. Forsgren, N., Humble, J., Kim, G., Brown, A., Kersten, N.: 2017 state of DevOps report. Puppetlabs (2017)
23. Velasquez, N.F., Kim, G., Kersten, N., Humble, J.: State of DevOps report 2018 (2018)
24. Kitchenham, B., Charters, S.: Guidelines for performing systematic literature reviews in software engineering version 2.3, vol. 45, no. 4ve (2007)
25. Genero, M., Cruz, J.A., Piattini, M.G.: Métodos de investigación en ingeniería del software (2014)
26. Solingen, R., Berghout, E.: The goal/question/metric method: a practical guide for quality improvement of software development (1999)
27. Santos, C.M.D.C, Pimenta, C.A.D.M, Nobre, M.R.C.: A estratégia PICO para a construção da pergunta de pesquisa e busca de evidências. Rev. Lat. Am. Enfermagem 15(3), 508–511 (2007). https://doi.org/10.1590/S0104-11692007000300023
28. Elberzhager, F., Münch, J., Nha, V.T.N.: A systematic mapping study on the combination of static and dynamic quality assurance techniques. Inf. Softw. Technol. 54(1), 1–15 (2012)
29. Cukier, D.: DevOps patterns to scale web applications using cloud services. In: Proceedings of the 2013 Companion Publication for Conference on Systems, Programming; Applications: Software for Humanity, vol. 38, pp. 143–152 (2013)
30. De Bayser, M., Azevedo, L.G., Cerqueira, R.: ResearchOps: the case for DevOps in scientific applications. In: Proceedings of the 2015 IFIP/IEEE International Symposium on Integrated Network Management, IM 2015, pp. 1398–1404 (2015)
31. Schneider, T.: Achieving cloud scalability with microservices and DevOps in the connected car domain. In: CEUR Workshop Proceedings, vol. 1559, pp. 138–141 (2016)
32. Gotimer, G., Stiehm, T.: DevOps advantages for testing: increasing quality through continuous delivery. CrossTalk Mag. 29(3), 13–18 (2016)
33. Jones, S., Noppen, J., Lettice, F.: Management challenges for DevOps adoption within UK SMEs. In: Proceedings of the 2nd International Workshop on Quality-Aware DevOps, pp. 7–11 (2016)
34. Nybom, K., Smeds, J., Porres, I.: On the impact of mixing responsibilities between Devs and Ops. In: Lecture Notes on Business Information Processing, vol. 251, pp. 131–143 (2016)
35. Lwakatare, L.E., et al.: Towards DevOps in the embedded systems domain: why is it so hard? In: Proceedings of the Annual Hawaii International Conference on System Sciences, pp. 5437–5446, March 2016
36. Ardulov, Y., Shchemelinin, D.: DevOps: reliability, monitoring and management with service asset and configuration management. In: imPACt 2017 - Internet, Mobile, Performance and Capacity, Cloud and Technology, November 2017
37. Elberzhager, F., Arif, T., Naab, M., Süß, I., Koban, S.: From agile development to DevOps: going towards faster releases at high quality - experiences from an industrial context. In: Lecture Notes on Business Information Processing, vol. 269, pp. 33–44 (2017)
38. Meirosu, C., John, W., Opsenica, M., Mecklin, T., Degirmenci, F., Dinsing, T.: Fueling the evolution. Ericsson Rev. (English Ed.), vol. 95, no. 2, pp. 70–81 (2017)
39. Perez-Palacin, D., Ridene, Y., Merseguer, J.: Quality assessment in DevOps: automated analysis of a tax fraud detection system. In: Proceedings of the 8th ACM/SPEC on International Conference on Performance Engineering Companion, pp. 133–138 (2017)
40. Laukkanen, E., Paasivaara, M., Itkonen, J., Lassenius, C., Arvonen, T.: Towards continuous delivery by reducing the feature freeze period: a case study. In: Proceedings of the 39th International Conference on Software Engineering: Software Engineering in Practice Track, pp. 23–32 (2017)

41. Fazal-Baqaie, M., Güldali, B., Oberthür, S.: Towards DevOps in multi-provider projects. In: CEUR Workshop Proceedings, vol. 1806, pp. 18–21 (2017)
42. Li, H., Chen, T.-H. P, Hassan, A.E., Nasser, M., Flora, P.: Adopting autonomic computing capabilities in existing large-scale systems: an industrial experience report. In: Proceedings of International Conference on Software Engineering, pp. 1–10 (2018)
43. Herger, L.M., Bodarky, M., Fonseca, C.A.: Breaking down the barriers for moving an enterprise to cloud. In: IEEE International Conference on Cloud Computing, CLOUD, pp. 572–576, July 2018
44. Zheng, E., Gates-Idem, P., Lavin, M.: Building a Virtually air-gapped secure environment in aws: with principles of DevOps security program and secure software delivery. In: Proceedings of the 5th Annual Symposium and Bootcamp on Hot Topics in the Science of Security, pp. 11:1–11:8 (2018)
45. Cleveland, S.B., Dooley, R., Perry, D., Stubbs, J., Fonner, J.M., Jacobs, G.A.: Building science gateway infrastructure in the middle of the pacific and beyond: experiences using the agave deployer and agave platform to build science gateways. In: ACM International Conference Proceeding Series (2018)
46. Rahman, A.: Characteristics of Defective infrastructure as code scripts in DevOps. In: Proceedings of the 40th International Conference on Software Engineering: Companion Proceedings, pp. 476–479 (2018)
47. Sampedro, Z., Holt, A., Hauser, T.: Continuous Integration and Delivery for HPC: using singularity and jenkins. In: Proceedings of the Practice and Experience on Advanced Research Computing, pp. 6:1–6:6 (2018)
48. Senapathi, M., Buchan, J., Osman, H.: DevOps capabilities, practices, and challenges: insights from a case study. In: ACM International Conference Proceeding Series, vol. Part F1377 (2018)
49. Düllmann, T.F., Paule, C., van Hoorn, A.: Exploiting DevOps practices for dependable and secure continuous delivery pipelines. In: Proceedings of the 4th International Workshop on Rapid Continuous Software Engineering, pp. 27–30 (2018)
50. Guija, D., Siddiqui, M.S.: Identity and Access control for micro-services based 5G NFV platforms. In: Proceedings of the 13th International Conference on Availability, Reliability and Security, pp. 46:1–46:10 (2018)
51. Ivanov, V., Smolander, K.: Implementation of a DevOps pipeline for serverless applications. In: Lecture Notes in Computer Science (Including subseries Lecture Notes in Artificial Intelligence and Lecture Notes in Bioinformatics) (2018)
52. Lewerentz, M., et al.: Implementing DevOps practices at the control and data acquisition system of an experimental fusion device. Fusion Eng. Des. **146**, Part A, 40–45 (2019). https://doi.org/10.1016/j.fusengdes.2018.11.022. ISSN 0920-3796
53. Lautenschlager, F., Ciolkowski, M.: Making runtime data useful for incident diagnosis: an experience report. In: Lecture Notes in Computer Science (Including subseries Lecture Notes in Artificial Intelligence and Lecture Notes in Bioinformatics), LNCS, vol. 11271, pp. 422–430 (2018)
54. Jacobsen, D.M., Kleinman, R., Longley, H.: Managing a Cray supercomputer as a git branch. Concurr. Comput. Pract. Experience **31**, e5092 (2019). https://doi.org/10.1002/cpe.5092
55. Bruza, M., Reith, M.: Teaming with silicon valley to enable multi-domain command and control. In: 13th International Conference on Cyber Warfare and Security, ICCWS 2018, pp. 663–667 (2018)
56. Albuquerque, A.B., Cruz, V.L.: Implementing DevOps in legacy systems. Adv. Intell. Syst. Comput. **860**, 143–161 (2019)
57. Savor, T., Douglas, M., Gentili, M., Williams, L., Beck, K., Stumm, M.: Continuous deployment at Facebook and OANDA. In: 2016 IEEE/ACM 38th International Conference on Software Engineering Companion (ICSE-C), pp. 21–30 (2016)

Reinforcing DevOps Generic Process with a Guidance Based on the Basic Profile of ISO/IEC 29110

Mirna Muñoz$^{(\boxtimes)}$ and Mario Negrete

Centro de Investigación en Matemáticas, Parque Quantum,
Ciudad del Conocimiento, Avenida Lassec,
Andador Galileo Galilei, Manzana, 3 Lote 7, 98160 Zacatecas, Mexico
{mirna.munoz,mario.negrete}@cimat.mx

Abstract. Nowadays customers are expecting faster feedback and changes related to issues or feature requests but providing a solution to this need, therefore organizations such as IBM, Facebook and Firefox are implementing DevOps practices. However, even when information about DevOps, such as principles, practices, suggestions or even a maturity or capability model, could be find in many sources containing, there is not a guidance on how to adopt and implement it. This fact becomes very important to achieve a reinforcement or evolution of the software development process and its implementation, according to the specific needs of an organization. As result of this lack of guidance, many authors have different understanding about what means the implementation of DevOps practices, causing confusion and an overwhelming situation among team within an organization regarding to the technology, processes or adoption of such practices. To provide a solution of the actual situation, this paper presents a guideline that allows the establishment of a generic DevOps process proposed, which aims to achieve the implementation of a DevOps approach reinforced with the proven practices of the Basic profile of the ISO/IEC 29110. Besides, it enables the organization to achieve a definition or improvement of process, roles and work products involved in a DevOps environment in an easy way.

Keywords: DevOps · Guidance · ISO/IEC 29110 · Software · Implementation process

1 Introduction

In recent years, users and customers of today's applications expect fast feedback to issues and feature requests [1]. In an effort to provide a solution, software organizations are trying to find new ways to achieve this need, such as the case of DevOps environment.

DevOps is a set of practices, which emerges to bridge the gaps between the operations team and developers enabling better collaboration [5].

© Springer Nature Switzerland AG 2020
J. Mejia et al. (Eds.): CIMPS 2019, AISC 1071, pp. 65–79, 2020.
https://doi.org/10.1007/978-3-030-33547-2_6

However, adapting DevOps practices to the current process is a big challenge that involves cultural and organizational aspects, bringing new concepts like automation in processes and tools, that most of the time overwhelming the team.

Moreover, the separation between developers and operations team is commonly found in many organizations today, causing one of the main obstacles for fast and frequent releases of applications [1].

One of the reasons that developers aim to push changes into production quickly, but the operations team aim to keep production environments stable, there are different goals and incompatible processes followed by two groups [2].

In this context, the productivity of some IT departments are increased because the agile response of DevOps culture, in which the life cycle is accelerated and best practices implemented in the current process, helps to achieve it [3].

Besides, a DevOps environment implies an automation in process and tools as shown in the results of an interview performed to 18 case organizations from various software domains, in which is mentioned that commodity tools such as version control, continuous integration, UI testing, performance, are used in almost every organization, which have implemented a DevOps environment [4].

Moreover, DevOps enables a continuously changing, arising new approaches, tools, and artifacts, so that it is hard to integrate and combine all these things in one way [1].

One of the main features of DevOps is the continuous delivery of software, in order to get it, it is necessary to beat organizational and cultural changes to eliminate the separation among team [6].

Taking as base the results obtained from a Systematic Literature Review (SLR) published in [8], in which were highlighted as main problems which are divided into 3 sections with 3 main problems. (1) process are not defined, lack of guidance and uncoordinated activities. This paper presents a proposal of guidance that allows an organization to implement a DevOps environment in an easy way.

The rest of the paper is organized as follows: in Sect. 2 Background, Sect. 3 Reinforced DevOps, Sect. 4 Guidance complete, Sect. 5 Discussion, Conclusions, and Next Steps.

2   Background

As first step to reinforce a DevOps process, a generic DevOps process was defined. Therefore, using the results provided by the SLR performed in [6] and the analysis of related works [10–13], the phases, activities and tasks of a DevOps approach were identified, analyzed and classified. The Table 1 shows the analysis performed to achieve it. Besides, Fig. 1 shows the generic DevOps process defined and next briefly described.

Table 1. DevOps elements sources.

Elements	The science of Lean software and DevOps accelerate [10]	The DevOps handbook [11]	DevOps a software architects perspective [12]	The Phoenix project, IT Revolution [13]
Phases			X	
Activities	X	X	X	X
Roles			X	
Work products	X	X	X	X
DevOps Principles		X		X
Tasks	X	X	X	X

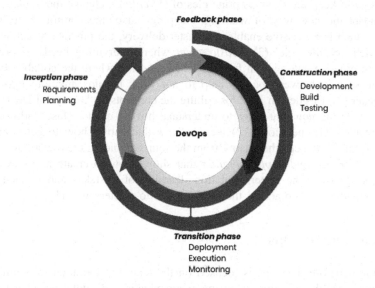

Fig. 1. Generic DevOps process.

As Fig. 1 shows, the generic process consists of 4 phases as next described:

- *Inception.* This phase aims to develop the project planning and define the requirements, the schedule, tools, and the team needed to perform the whole project. This phase has 2 activities: *requirements* and *planning* [10–13].
- *Construction.* This phase aims to handle the pipeline which is: commit to the stage, approve changes, testing in different ways (e.g. smoke test, exploratory testing, unit test, modular test, regression test) to review software components, which will be developed in the production. The main intention is to automate the activities following the guidance in order to get software features. This phase has 3 activities: *development, build* and *test* [10–13].
- *Transition.* This phase aims to ship the software in a server or the necessary hardware to run it. It might be using a manual or an automated deliverance. After

having implemented the software, the release part consists of creating the software deliverables based on new features added or a set of fixed bugs of existing features. Then, the software is shipped to the customer to be used. Next, the operation team keeps monitoring the servers or applications in order to alert about possible issues. This phase has 3 activities: *deployment, execution,* and *monitoring* [10–13].

- *Feedback.* This phase aims to provide knowledge about how to deal with the issues and to improve the current process [12]. To achieve it, the stakeholders and work team give a feedback to prevent any issue and maintain the software working. It is important to highlight that the generic DevOps process has 4 phases, 8 activities and 40 tasks, which have to be followed in order to perform a software development process with DevOps culture.

The DevOps process is based on Lean software development and XP agile methodology to keep the DevOps principles of (1) *culture,* change the thinking of the team to adapt the new way of working, stepping out of the comfort zone, (2) *automation,* which is a massive enabler for faster delivery, the practice of automate the software delivery lifecycle, (3) *measurement* where everything needs to provide a feedback in order to automate in future, and (4) *sharing* where the collaboration and knowledge sharing between people help to not work in silos anymore making the process better [7]. Besides, in DevOps culture the concepts or "3 ways of DevOps" are highlighted: (1) the *flow* that helps to understand how to create a fast workflow from development for IT operations; (2) the *feedback* that shows how to get continuous amplified feedback to keep the quality from the source avoiding the rework; and (3) the *continuous learning and implementation* that shows how to create a culture, which simultaneously foster the experimentation, learning of mistakes and understanding, being the repetition and practice, the prerequisites to mastery it [11].

3 Reinforced DevOps

As mentioned before, DevOps is focused on the team collaboration to automate the process using visible workflows, but many organizations do not want to bet on this approach because it implies risks or specific issues that can be classified within 3 categories: guidance, processes, and team [8]. Figure 2 shows the specific issues that affect a DevOps environment and how can they be improved with ISO/IEC 29110. However, the DevOps process could be improved with ISO/IEC 29110 series that could ensure the quality product and reduce the rework, while providing guidance that

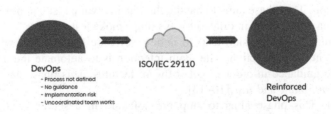

Fig. 2. DevOps transformation.

enables a process definition, the risk management and providing roles to keep a self-coordinate team [8]. Case studies about organizations in some countries with similar problems as detected in DevOps environments were solved by using artifacts, roles, tasks of the Basic profile of ISO/IEC 29110 [9].

The proposal of the guidance is to use practices, tasks, work products and roles of the Basic profile of ISO/IEC 29110 series, in order to get a reinforced DevOps approach to ensure the quality, avoiding rework and reducing risks.

The method to create the guidance is: (1) identify generic DevOps phases and classify information from many sources; (2) classify the phases, activities, tasks, techniques and tools of DevOps; (3) identify the issues to set up a DevOps process; (4) create a comparative table to check the tasks of Basic profile of the ISO/IEC 29110 needed to improve the generic DevOps process; and (5) build the reinforced DevOps process implementing the selected tasks of the Basic profile of the ISO/IEC 29110.

The structure of reinforced DevOps keeps the 4 phases: Inception, Construction, Transition, and Feedback. Figure 3 shows an overview of the reinforced DevOps process as well as the activities of the Basic profile of ISO/IEC 29110 that are used to reinforce each phase.

The reinforced DevOps process was defined using process elements such as phases, activities, tasks, inputs and output products and roles.

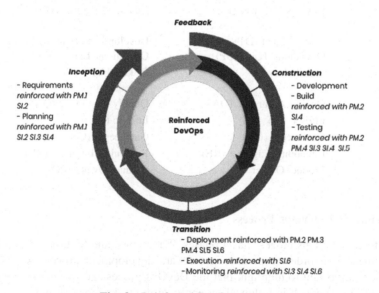

Fig. 3. Reinforced DevOps process.

3.1   Reinforced DevOps Roles

To select the roles for the reinforced DevOps, both roles, from ISO/IEC 29110 and the generic DevOps process, were taken into account to develop a comparative matrix. The matrix allows reviewing the activities performed by each role so that, it was possible to identify the relationship among them.

To select the roles next criteria was applied. If a DevOps role doesn't have a relationship with any ISO/IEC 29110 role, the role is considered as part of the reinforced DevOps to accomplish their responsibility.

In the case of the work team, were divided in two groups, development and operation team, every team have different activities but they should work together.

Table 2 shows the selection of reinforced DevOps roles. It is important to mention that a person can cover more than 1 role in the organization.

Table 2. DevOps roles adaptation.

ISO/IEC 29110	DevOps	Reinforced DevOps
Analyst (AN)	Value Stream Manager (VSM)	Analyst (AN)
Project Manager (PM)	Process Master (PM)	Process Master (PM)
Technical Leader (TL)	DevOps Engineer (DE)	DevOps Engineer (DE)
Work Team (WT)	Development (DT) and Operations Team (OP)	Development Team (DT) and Operations Team (OP)
Customer (CUS)	Stakeholder	Stakeholder
	Quality Assurance (QA)	Quality Assurance (QA)
	Infosec (IS)	Infosec (IS)
	Gatekeeper (GK)	Gatekeeper (GK)
	Reliability Engineer (RE)	Reliability Engineer (RE)
	Product Owner	Product Owner (PO)

3.2   Reinforced DevOps Process

The reinforced DevOps process has 4 phases, 9 activities, and 87 tasks, which have to be implemented in order to perform a software development process with DevOps culture. It is important to highlight that the DevOps process keeps an incremental and iterative lifecycle, which is followed by applying the tasks using sprints.

The reinforced phases are next briefly described:

1. *Inception phase*: this phase aims to define the requirements, which are the initial part of software development. As Fig. 4 shows it has 2 activities (requirement and planning).

Activity	Role	Tasks
	PM, AN	I1.1 Review the Statement of Work or the contract which specify the understanding of the project.
	PO, PM	I.1.2 Define with the stakeholders the delivery instructions of each one of the deliverables specified in the statement of work.
	PO, PM, AN	I.1.3 Document or update the user stories. Identify and consult information sources such as stakeholders (interviews), previous systems, documents, etc. Analyze the identified user stories for determining its feasibility and the project scope. generate or update the product backlog.
	PM, AN	I.1.4 Identify the specific tasks and setup tasks to be performed in order to produce the deliverables and the software components identified in the statement of work or contract. Identify the tasks to perform the delivery Instructions and document the tasks.
	PO, AN	I.1.5 Perform the backlog refinement session if it's necessary to define what user stories are expected in the next sprint and communicate uncertainties for the user stories that are unclear. The product owner can improve the user stories where needed.
I1. Requirements	PM, AN	I.1.6 Identify and document the Resources: human, material, equipment and tools, standards, including the required training of the Work Team to perform the project. Include in the schedule the dates when Resources and training will be needed.
	PM, AN	I.1.7 Establish the Composition of Work Team assigning roles and responsibilities according to the Resources in order to define the hierarchical organization oriented to value stream market.
	PM, DT, QA, IS, GK, AN, RE, DE	I.1.8 Define the complexity and time of the user stories using an estimation technique among the team. It's recommended to use techniques like planning pocket to estimate
	PM AN	I.1.9 Assign estimated start and completion dates to each one of the product backlog tasks in order to create the schedule of the project tasks taking into account the assigned resources, unplanned work, technical debt, sequence and dependency of the tasks. Besides, it's recommended to agree on a shared goal among the whole team to achieve in the sprint.
	PO, PM	I.1.10 Calculate and document the project estimated effort and cost using function points.
	PM, AN	I.1.11 Verify and obtain approval of the product backlog. Verify the correctness and testability of the product backlog and its consistency with the product description. Additionally, review that user stories are complete, unambiguous and not contradictory with the acceptance criteria. The results found are documented in verification results and corrections are made until the product backlog is approved by the analyst. If significant changes were needed, initiate a change request.
	PM, DT, QA, OT, IS, GK, AN, RE, DE	I.2.1 Reduce the technique debt and unplanned work taking into consideration the tasks in the current sprint. If the impediment board has pending tickets.
	PM, AN	I.2.2 Identify and document the risks which may affect the project.
	PM, AN	I.2.3 Define the version control among the team in every component and deliverable using version control technique. It's strongly recommended apply a nomenclature of phase, major and minor changes during the project in the project plan.
I2. Planning	PM, AN	I.2.4 Generate the project plan integrating the elements previously identified and documented.
	PM, AN	I.2.5 Include product description, scope, objectives, and deliverables in the project plan.
	PO, PM, DT, QA, OT, IS, GK, AN, RE, DE	I.2.6 Verify the current project plan with the team members in order to achieve a common understanding and get their engagement with the project to obtain approval of the project plan. The results found are documented in a verification results and corrections are made until the document is approved by process master.
	PM, AM	I.2.7 Document the preliminary version of the software user documentation or update the existing one, if appropriate.
	PM, AN	I.2.8 Verify and obtain approval of the software user documentation, if appropriate. Verify consistency of the software user documentation with the product backlog. The results found are documented in verification results and corrections are made until the

Fig. 4. Inception phase activities and tasks

Activity	Role	Tasks
C1. Development	DE	C.1.1 Generate configuration files if it's necessary to apply in server, containers, and tools. It's strongly recommended to create templates according the project in order to use defined templates to save time in the configuration.
	DE	C.1.2 Create any other script or configuration information required to create the software infrastructure.
	DE	C.1.3 Script the database migrations and seeders needed to create the database or update if necessary.
	DE	C.1.4 Run the scripts to keep the database up to date.
	DE, OP	C.1.5 Define trunk-based development using different environments and working in small batches. It's strongly recommended to use different branches to get better control, such as master, develop, and test as a minimum.
	DT, OT, DE	C.1.6 Establish an environment, which supports continuous integration. It's strongly recommended to use a Continuous integration server to define the configuration in a specific file. Add a way to prevent issues, recovering of service failure without delay.
	DT, OT, DE	C.1.7 Establish an environment, which supports deployment automation. It's strongly recommended to use a Continuous integration server to define the configuration in a specific file.
	DT, OT, DE	C.1.8 Establish an environment, which supports delivery automation. It's strongly recommended to use a Continuous integration server to define the configuration in a specific file.
	DT, OT, DE	C.1.9 Establish an environment, which supports the release automation. It's strongly recommended to use a Continuous integration server to define the configuration in a specific file.
C.2 Build	DE, OP	C.2.1 Create pre-configured virtual machine images or containers using the configuration files if exist. It's strongly recommended to use containers to keep the same configuration in all environment in whichever device and not to change or introduce unknown variables in the current configuration.
	PM	2.2 Ensure that every environment has the same configuration. It's strongly recommended to have a virtual machine in order to reduce setup time.
	DE, OP	C.2.3 Define the security strategy to apply in containers or the environment needed. It's recommended to use techniques such as middleware, two-factor authentication or any technique needed.
	DE, OP	C.2.4 Set or update the security strategy into the environment to use in the project.
	DT, OT	C.2.5 Construct or update software components based on the detailed part of the software design and following the acceptance criteria
	PM, AN	C.2.6 Monitor the project plan execution and record actual data in progress status record.
	PO, PM, AN	C.2.7 Conduct revision meetings with the stakeholder, record agreements and track them to closure. Change request initiated by the stakeholder or initiated by the work team, which affects the stakeholder, needs to be negotiated to reach acceptance of both parties. If necessary, update the project plan according to the new agreement with stakeholders.
	PO, PM, AN	C.2.8 Define a change approval process analyzing and evaluating the change request in cost, schedule, and technical impact. The change request can be initiated externally by the stakeholder or internally by the team. Update the project plan, if the accepted change does not affect agreements with stakeholders. change request, which affects those agreements, needs to be negotiated by both parties and will be added to the product backlog to be considered in the next sprint.
	PM	C.2.9 Conduct revision meetings with the QA, Infosec, analyst, reliability engineer, work team. Identify problems, review risk status, record agreements and track them to closure.

Fig. 5. Construction phase activities and tasks.

C.3 Testing	DT	C.3.1 Establish or update test cases and test procedures for integration testing and add them in the testing automation based on the product backlog and software design. Stakeholder provides testing data if needed.
	PM, AN	C.3.2 Verify and obtain approval of the test cases and test procedures. Verify consistency among product backlog, software design, and test cases and test procedures. The results found are documented in verification results and corrections are made until the document is approved by the analyst.
	DT	C.3.3 Understand test cases and test procedures. Set or update the testing environment.
	AN	C.3.4 Update the traceability record incorporating the test cases and test procedures.
	PM	C.3.5 Incorporate the software design and traceability record to the software configuration as part of the baseline. Incorporate the test cases, and test procedures to the project repository.
	DT	C.3.6 Design or update unit test cases and apply them to verify that the software components implement the detailed part of the software design.
	DT, DE	C.3.7 Establish an environment, which supports testing automation. It's strongly recommended to use a Continuous integration server to define the configuration in a specific file.
	DT OT	C.3.8 Integrate the software using software components and update test cases and test procedures for integration testing, as needed.
	DT AN	C.3.9 Perform software tests using test cases and test procedures for integration and document results in the test report.
	DT	C.3.10 Correct the defects or run the testing configuration found until successful unit test (reaching exit criteria) is achieved. It's strongly recommended to get a record of the tests performed and results even if is a failed result.
	DT, QA, OT, IS, GK, DE	C.3.11 Update the traceability record incorporating software components constructed or modified.
	DT	C.3.12 Incorporate software components and traceability record to the software configuration as part of the baseline.
	DT, AN	C.3.13 Update the traceability record, if appropriate.
	AN	C.3.14 Incorporate the test cases and test procedures, software, traceability record, test report, product operation guide and software user documentation to the software configuration as part of the baseline.

Fig. 5. (*continued*)

2. *Construction phase*: this phase aims to define configuration files needed to automate the software implementation process, it also defines the pipeline to deliver the application after applying the test cases. As Fig. 5 shows it has 3 activities (development, build and test).
3. *Transition phase*: this phase aims to define strategies to recover after any issue, evaluate the project progress and monitoring the application performance. As Fig. 6 shows it has 3 activities (deployment, execution, and monitoring).
4. *Feedback phase*: this phase aims to get a continuous feedback about team, customer and project (See Fig. 7).

Activity	Role	Tasks
T.1 Deployment	RE	T.1.1 Perform backup according to the version control strategy. If necessary, consider changing the external providers in order to ensure the backup.
	RE	T.1.2 Restore the previous branch of production. If it's necessary using a rollback if an issue is detected.
	RE	T.1.3 Setup in the continuous integration file a way to recover from service failure without delay, if necessary.
	PM	T.1.4 Evaluate project progress with respect to the project plan, comparing: - actual tasks against planned tasks - actual results against established project goals - actual resource allocation against planned resources - actual cost against budget estimated - review the burndown chart with the forecast - actual risk against previously identified
	PM	T.1.5 Establish actions to correct deviations, issues, unplanned work and technical debt and identified risks concerning the accomplishment of the plan, as needed, document them in correction register and track them to closure.
	PM	T.1.6 Identify changes to user stories and/or project plan to address major deviations, potential risks or problems concerning the accomplishment of the plan, document them in change request and track them to closure.
	DT, DE	T.1.7 Update project repository and cloud services.
	AN	T.1.8 Document the product operation guide or update the current guide, if appropriate.
	PM, AN	T.1.9 Verify and obtain approval of the product operation guide, if appropriate, verify consistency of the product operation guide with the software. the results found are documented in a verification results and corrections are made until the document is approved by designer.
	PM, AN	T.1.10 Document the software user documentation or update the current one, if appropriate.
	PM, AN	T.1.11 Verify and obtain approval of the software user documentation, if appropriate, verify the consistency of the software user documentation with the software. The results found are documented in verification results and corrections are made until the document is approved by stakeholder.
	AN	T.1.12 Document the maintenance documentation or update the current one.
	PM, AN	T.1.13 Verify and obtain approval of the maintenance documentation. verify the consistency of maintenance documentation with software configuration. the results found are documented in verification results and corrections are made until the document is approved by DevOps engineer.
T.2 Execution	PM, AN	T.2.1 Incorporate the maintenance documentation as baseline for the software configuration.
	PO, PM	T.2.2 Perform delivery according to delivery instructions in the continuous integration file.
	DT	T.2.3 Package the code in ways suitable for deployment.
	DE, RE	T.2.4 Deploy the code in the suitable server to check the functions. It's strongly recommended to apply the same process.
	DE	T.2.5 Copy packages or files into production servers or in the place needed depending on the project target. It's strongly recommended to have a deployment automation file to avoid this task.
T.3 Monitoring	DE	T.3.1 Decide the monitoring tools to apply in the server. It's strongly recommended to use a monitoring tool to get graphs about the project deployment.
	GK	T.3.2 Ensure the release performance checking the monitoring graphics.
	GK	T.3.3 Measure the statistics of the monitoring tool of the server.
	GK	T.3.4 Configure proactive notification according the project needed to have a log of issues. It's strongly recommended to get information about features like security, availability, performance, etc.

Fig. 6. Transition phase activities and tasks.

Activity	Role	Tasks
F.1 Feedback	PM	F.1.1 Evaluate the team after finishing the sprint to know what went well and what went not so well and what could be improved.
	PM	F.1.2 Analyze the organizational level of the organization although Westrum organizational which consist in classify in pathological, bureaucratic, generative according to the Westrum survey to know the job satisfaction.
	PM	F. 1.3 Get continuous user behavior in the application using the monitoring to get feedback in order to get future improvement.
	PM	F.1.4 Publish post-mortems with the information about issues, things to improve as widely possible and do visible for the whole team.

Fig. 7. Feedback phase activities and tasks.

4 Guidance Implementation

There are a total of 87 tasks in a reinforced DevOps process among the four phases (inception, construction, transition and feedback), therefore, it could become too complicated for organizations to adopt or implement all practices at one time, for the next reasons: (1) lack of technology knowledge, (2) lack of techniques and practices knowledge; (3) lack of experience implementing a process and (4) lack of experience improving a process and (5) lack of budget.

In order to facilitate the implementation of the reinforced DevOps process, a strategy was established according to the Maslow pyramid hierarchy of needs [16], which is a theory in psychology.

According Maslow pyramid, the higher needs in this hierarchy only come into focus when the lower needs in the pyramid are satisfied [14].

Figure 8 shows the DevOps principles adapted to the Maslow pyramid as a strategy to solve the needs for implementing a DevOps in a right way. The 4 principles of DevOps (flow, feedback and continual learning, and experimentation) were divided into a hierarchy of needs to build the guidance, which should be covered in order to achieve the right implementation of the reinforced DevOps.

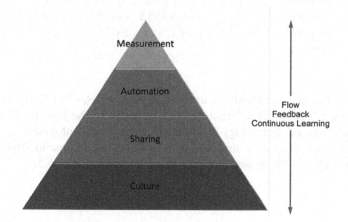

Fig. 8. DevOps principles adapted to the Maslow's pyramid

Our hypothesis is that by this way, it is possible to reduce the risk of overwhelming a team, if they don't have enough experience to change.

The first principle is *culture*, to achieve this principle 49 tasks should be implemented. The benefits achieving this principle are: the information is sharing active, messengers are not punished when they deliver news of failures or other bad news, responsibilities are shared, cross-functional collaboration among the team is encouraged and rewarded, new ideas are welcomed and failures are treated primarily as opportunities to improve the system. The tasks included in the Culture principle are shown in Table 3.

Table 3. Reinforced DevOps practices to be implemented to achieve the culture principle.

Phase	Tasks
Inception	1.1, 1.2 1.3, 1.4, 1.6, 1.7, 1.8, 1.9, 1.10, 2.1, 2.2, 2.3, 2.4, 2.5, 2.7, 2.13, 2.14, 2.15
Construction	1.1, 1.2, 1.3,1.4, 1.5, 2.1, 2.3, 2.4, 2.5, 2.6, 2.7, 2.8, 3.1, 3.6, 3.8, 3.9, 3.10
Transition	1.1, 1.4, 1.5, 1.6, 1.7, 1.8, 2.2, 2.3, 2.4, 2.5, 3.5
Feedback	1.1, 1.2, 1.4

The second principle is *sharing*, to achieve this principle, 25 should be added about verification and validation actions. The benefits of implementing this are: the organization understands what they are doing, the team understands each other's goals and priorities, the knowledge is shared across the team, the members of the team are collaborating with each other, the team verifies work products and validate deliverables with the stakeholders. The tasks included in the sharing principle are shown in Table 4:

Table 4. Reinforced DevOps practices to be implemented to achieve the sharing principle.

Phase	Tasks
Inception	1.11, 2.6, 2.8, 2.9, 2.10, 2.11, 2.12, 2.16, 2.17
Construction	2.2, 2.9, 3.2, 3.3, 3.4, 3.5, 3.11, 3.12, 3.13, 3.14
Transition	1.9, 1.10, 1.11, 1.12, 1.13, 2.1
Feedback	

The third principle is *automation*, to achieve this principle, 7 tasks should be added about continuous actions in different parts of the project. The benefits of this implementation are: shared technology knowledge, automated deployment and support, reduce testing time, keep up with monitoring data and reduce time to setup functions. The tasks included in the automation principle are shown in Table 5.

Table 5. Automation process.

Phase	Tasks
Inception	1.5
Construction	1.6, 1.7, 1.8, 1.9, 3.7
Transition	1.2
Feedback	

The fourth principle is *measurement*, to achieve this principle, 6 tasks should be added about feedback. The benefits of this implementation are: reduce rework, increase quality and to be a candidate to get a certification of the Basic profile of ISO/IEC 29110. The tasks included in the automation principle are shown in Table 6:

Table 6. Measurement process.

Phase	Tasks
Inception	
Construction	
Transition	1.3, 3.1, 3.2, 3.3, 3.4
Feedback	1.3

5 Discussion

Nowadays, DevOps is everywhere or so it seems, as one of the most ubiquitous trends in software development, its adoption has been surrounded by promises about faster roll-outs, agility, responsiveness, and gains in efficiency using automation in processes [15]. Development and Operation team that is in an environment of collaboration with continuous integration/delivery is able to deliver the software, which will be more aligned with business need.

DevOps is related with Continuous Software Engineering (CSE), which is a new approach, which aims to establish connections between software engineering activities to accelerate and increase the efficiency of the software process [16].

DevOps integrates many techniques in order to decrease the time between changing a system as well as the production environment, using monitoring practices to ensure quality and get feedbacks [17].

However, to achieve the right implementation of it, depends on the expertise, experience, and its correct understanding.

Moreover, the need of a guidance for a right understanding and implementation of DevOps was highlighted in [8] is solved with the proposal of guidance presented in this paper.

6 Conclusions and Next Steps

DevOps provides a cultural change in organizations that uses:

(1) a set of practices to get a visible workflow, which can be performed by the whole team; (2) the automation of processes, which cause a continuous feedback during the project and; (3) the improvement of learning through that experimentation.

With the implementation of the Reinforced DevOps is expected that the DevOps features such as: project control, process definition guidance, communication improvement, and reduction of risks, while avoiding reworking by applying verification and validation tasks, can be implemented easily in any organization.

Table 7 shows the benefits acquired using the guidance.

Table 7. DevOps features.

DevOps	Reinforced DevOps
Flow	Flow
Feedback	Feedback
Continual learning and experimentation	Continual learning and experimentation
Culture	Culture
Automation	Automation
Measurement	Measurement
Sharing	Sharing
	Control
	Guidance
	Process
	Communication
	Risk reduction
	Avoid reworking
	Verification
	Validation
	Candidate of certification in ISO/IEC 29110 Basic profile

The main limitation of this proposal is for non-critical software and doesn't consider specific security practices. The next step in this research work is to create a web platform, having the next functionalities: (1) a survey to verify the state of the organization in order to tailor the guidance; (2) Guidance, that are based on specific needs, which provides the enough elements to adapt the reinforced DevOps in an organization depending on its maturity level; (3) a set of work products are explained as well as filled templates and; (4) Roles definition and their responsibilities.

References

1. Wettinger, J., Breitenbücher, U., Kopp, O., Leymann, F.: Streamlining DevOps automation for cloud applications using TOSCA as standardized metamodel **56**, 317–332 (2016)
2. Hüttermann, M.: DevOps for Developers. Apress, New York (2012)
3. Muñoz, M., Díaz, O.: Engineering and Management of Data Centers (2017)
4. Gupta, V., Kapur, P.J., Kumar, D.: Modeling and measuring attributes influencing DevOps implementation in an enterprise using structural equation modeling. Inf. Softw. Technol. **92**, 75–91 (2017)
5. Humble, J., Farley, D.: Continuous Delivery: Reliable Software Releases Through Build, Test, and Deployment Automation. Addison-Wesley Professional, Boston (2010)
6. Nelson-Smith, S.: Test-Driven Infrastructure with Chef. O'Reilly Media Inc., Sebastopol (2013)
7. Kaiser, A.K.: Introduction to DevOps (2018)
8. Muñoz, M., Negrete, M., Mejía, J.: Proposal to avoid issues in the DevOps implementation: a systematic literature review (2019)
9. Laporte, C., García, L., Bruggman, M.: Implementation and Certification of ISO/IEC 29110 in an IT Startup in Peru (2015)
10. Nicole, F., Jez, H., Gene, K.: The Science of Lean Software and DevOps Accelerate, Building and Scaling High Performing Technology Organizations. IT Revolution, Portland (2018)
11. Gene, K., Jez, H., Patrick, D., John, W.: The DevOps Handbook. How to Create World-Class Agility, Reliability & Security in Technology Organizations. IT Revolution, Portland (2016)
12. Bass, L., Weber, I., Zhu, L.: DevOps A Software Architects Perspective. Addison Wesley, Boston (2015). SEI
13. Gene, K., Kevin, B., George, S.: The Phoenix Project. IT Revolution, Portland (2013)
14. Abulof, U.: Introduction: why we need maslow in the twenty-first century 508–509 (2017)
15. Steve, M.: DevOps finding room for security. Netw. Secur. **2018**(7), 15–20 (2018)
16. Sánchez-Gordón, M., Colomo-Palacios, R.: Characterizing DevOps culture: a systematic literature review (2018)
17. Balalaie, A., Heydarnoori, A., Jamshidi, P.: Microservices architecture enables DevOps: migration to a cloud-native architecture. IEEE Softw. **33**, 42–52 (2016)

A Selection Process of Graph Databases Based on Business Requirements

Víctor Ortega[✉], Leobardo Ruiz, Luis Gutierrez, and Francisco Cervantes

Department of Electronics, Systems and Information Technology, ITESO Jesuit University of Guadalajara, 45604 Tlaquepaque, Mexico
{vortega,ms716695,lgutierrez,fcervantes}@iteso.mx

Abstract. Several graph databases provide support to analyze a large amount of highly connected data, and it is not trivial for a company to choose the right one. We propose a new process that allows analysts to select the database that suits best to the business requirements. The proposed selection process makes possible to benchmark several graph databases according to the user needs by considering metrics such as querying capabilities, built-in functions, performance analysis, and user experience. We have selected some of the most popular native graph database engines to test our approach to solve a given problem. Our proposed selection process has been useful to design benchmarks and provides valuable information to decide which graph database to choose. The presented approach can be easily applied to a wide number of applications such as social network, market basket analysis, fraud detection, and others.

Keywords: Graph databases · Benchmarking · Selection process

1 Introduction

In the last decade, the large amount of information generated from multiple data sources such as social networks and mobile devices has led the relational database management systems to their limit. As larger the dataset, it becomes more difficult to process using traditional data processing applications like relational database management systems and data warehousing tools. The challenges include analysis, capture, curation, search, sharing, storage, transfer, and visualization [1–3]. This fact has driven to look for alternatives such as non-relational databases. NoSQL databases provide several options for storing, accessing and manipulating data at big scale; examples are key-value, document-based, column-based, and graph-based stores, all of these improves the querying performance that relational SQL databases have on large and unstructured datasets [4].

One of the most promising approaches is the Graph Database (GDB). This kind of databases includes support for an expressive graph data model with heterogeneous vertices and edges, powerful query and graph mining capabilities, ease of use, as well as high performance and scalability [5]. All these features are needed to manipulate and study highly connected data points. Other advantages of graph database systems include but not limited to the following topics: high-performance query capabilities on

J. Mejia et al. (Eds.): CIMPS 2019, AISC 1071, pp. 80–90, 2020.
https://doi.org/10.1007/978-3-030-33547-2_7

large datasets, data storage support in the order of petabytes (10^{15}), intuitive query language, appropriate for agile development, support of new types of data and suitable for irregular data, and optimized for data mining operations [6]. Overall, these features can be classified into three major areas: performance, flexibility, and agility.

Performance is a relevant factor when it comes to data relationship handling. Graph databases massively improve the performance when dealing with network examination and depth-based queries traversing out from a selected starting set of nodes within a graph database. They outperform the needed depth of join operations, which traditional relational databases implement in languages such as SQL when they need to go into a complex network to find how data relates to each other.

Flexibility plays an important role; it can allow data architects to modify database schemas, which can quickly adapt to significant changes because of new business needs. Instead of modeling a domain in advance, when new data need to be included in the graph structure, the schema can be updated on the fly as data are pushed to the graph.

Agility is highly appreciated due to most software companies adopting work methodologies such as Scrum and Kanban. This graph database feature aligns with such lean interactive development practices, which lets the graph evolve with the rest of the application and evolving business needs.

From an operational perspective, there are three types of graph database: true graph databases, triple stores (also named RDF stores), and conventional databases that provide some graph capabilities. True graph databases support index-free adjacency, which allows graph traversals without needing an index, while triple stores require indexing to perform traversals [7, 8]. True graph databases are designed to support property graphs where these properties are applied either to vertices or edges, and recently some triple stores have added this capability.

Most of the graph databases support a unique main query language, which allows us to include scenarios such as language type, ease of usage, SQL similarity [9], and evaluate graph data and queries portability across other platforms.

So far, we have described the benefits of using graph databases to address certain types of problems where it is necessary to analyze large amounts of highly related data. However, for companies, it is not trivial to identify which GDB is better according to their needs. Thus, the selection of a GDB becomes a challenge that can impact at an operational level and even at an economic level.

In this work, we propose a new selection process for guiding analysts to select a suitable GDB to satisfy the business needs. The selection process flow consists of five main stages: the problem analysis, requirements analysis, GDB analysis, benchmarking, and GDB selection.

Our contributions are (1) a new selection process to select a graph database aligned to business requirements, (2) a detailed description of the proposed selection process in a particular scenario.

We organized the rest of this document as follows: Sect. 2 describes how other studies benchmark graph databases. Section 3 presents and describes in detail our proposed selection process to choose a graph database. Section 4 shows the application of the selection process on a case study with two datasets and their results. Finally, in Sect. 5, we present our conclusions.

2 Related Work

There are several commercial GDB engines that enterprises and organizations could use. However, choosing the best option is not trivial. Several authors as Batra et al. [10] and Nayak et al. [11] have described the advantages of NoSQL databases and how those are solving common issues on legacy systems which rely on Relational Databases (RDB). These works provide an overall description of what graph databases are offering as a solution and remark the advantages and disadvantages of each type of systems.

Batra in [10] performs a comparative between MySQL and Neo4j. It provides the following evaluation parameters: level of support/maturity to know how well tested the system is, security to understand what mechanisms are available to manage restrictions and multiple users, flexibility which allows system designers to extended new features through time-based on schema changes, and performance-based upon predefined queries on a sample database. The obtained results are that both systems performed well on the predefined queries with slight advantages for Neo4j, the graph databases are more flexible than relational databases without the need to restructure the schemas.

Nayak in [11] describes each type of NoSQL database type. These are classified into five categories, one of them being Graph databases, the categories are key-value store databases such as Amazon DynamoDB and RIAK, column-oriented databases like Big Table and Cassandra, document store databases using MongoDB and CouchDB as examples, and graph Databases where Neo4j is the only analyzed system. No specific criteria were used to compare features. However, the paper provides the advantages and disadvantages of each of the data stores. Some of the advantages of NoSQL over Relational are that they provide a wide range of data models to choose, easily scalable, DB admins are not required, some provide handlers to react to hardware failures, they are faster, more efficient and flexible. The disadvantages of NoSQL over Relational include that they are immature in most of the cases, no standard query language; some of them are not ACID compliant, no standard interface, maintenance is difficult. Nayak provides a proper classification of NoSQL databases and Batra a usefully comparative between an RDB and Graph database, but any of them provide a formal process to select a graph database aligned to business requirements.

There are other works, which have compared only graph database systems [12–14]. The first one is providing a high-level description of AllegroGraph, ArangoDB, InfiniteGraph, Neo4j, and OrientDB [12]. The criteria, considered in work use the following elements: Schema flexibility, Query Language, Sharding (ability to break up large datasets and distribute it across several replicated shards), Backups, Multi-model, Multi-architecture, Scalability, and Cloud Ready. Each of the elements is graded for all the databases with a score which can be Great, Good, Average, Bad or Does not support. The obtained results are just remarking that there is no perfect graph database for any type of problem. An important element which is not evaluated in this work is the performance on graph operations through the use of their supported query languages.

The other works comparing only graph databases that go beyond the high-level features and evaluate performance tend to have the support of a specific development group or vendor. One of them is a post from Dgraph comparing the loading data capacities and querying performance against Neo4j [13], the dataset is in RDF format, which is not supported natively by Neo4j. Therefore, it had to be loaded through data format conversions. Then a set of specific queries are executed on each database where Dgraph outperforms Neo4j in most cases. The other graph comparison work is benchmarking TigerGraph, Neo4j, Neptune, JanusGraph, and ArangoDB [14], the work is developed by the creators of TigerGraph, the comparison includes data loading times, storage size, and the response of graph queries for full graph or subsets of it. It uses a couple of graphs, one which is considered a small size and then a large dataset. The conclusion favors TigerGraph in most of the compared features claiming that their development is based on a natural, real-time, and Massively Parallel Processing platform approach with a high-level language introduced by the platform. Even though the comparison presents some useful metrics, it does not provide a guide to choose the suitable GDB according to the business requirements.

As far as we know, in the literature, there is not a transparent process for the election of a GDB driven by the problem, goals, requirements, and constraints. Our work has no intention of favoring any of the evaluated systems, and it provides a guide to choose the suitable GDB according to business needs.

Nonetheless, some works focused on selection processes for other types of software tools such the proposed by Maxville et al. [15], where they implement a framework for the selection of open source software from the repositories. The context-driven component evaluation framework (CdCE) uses a specific set of templates, criteria, ontologies, and data representations, with classifiers and test generators for filtering and evaluation. The proposal has three main phases: filtering, evaluation, and ranking items.

Lee et al. [16] developed a tool based on the Analytic Hierarchy Process (AHP) to adopt open-source software. AHP is a decision-making method that includes qualitative and quantitative techniques to support decisions. Their selection process has four phases: (1) identify the goal/project, selection criteria and product alternatives, (2) justify judgment matrix for each criterion, (3) justify judgment matrix for the alternative products, and (4) results for the final ranking and decision making.

Lourenço et al. [1] presented a qualitative evaluation of NoSQL databases and described the main characteristics of each type of databases based on the literature review. This paper focused on performance requirements of the databases and quality attributes like availability, consistency, durability, maintainability, read and write performance, reliability, robustness, scalability, stabilization time, and recovery. Finally, Lee presents a timeline of the evolution of NoSQL databases.

The last works above provide a set of stages, metrics, and use cases as a guideline to choose diverse kind of technology, but are not focused on graph databases. Therefore, it is required a well-defined selection process of graph databases based on business requirements.

3   Selection Process of GDB

In this section, we describe a new selection process that guide analysts to select a GDB that satisfies the business needs. The selection process flow consists of five main stages: problem analysis, requirements analysis, GDB analysis, benchmarking, and GDB selection (Fig. 1).

3.1   Problem Analysis (Stage 1)

The goal of this stage is to define the problem that must be solved using a GDB. This stage has two steps: the problem and goals definition. We assume that, in general, the problem consists of analyzing a large amount of highly connected data.

In the problem definition step (step 1.1), it must be described the business needs in terms of the GDB scope. In this step, the analyst should answer the following question: why do I need a GDB?

Once the analyst defined the problem, he must determine a set of goals in terms of the capabilities that a GDB can offer, for example, the supported amount of processing data, the kind of analysis to perform on the connected data, filters, dynamic analysis, graph algorithms, and memory.

3.2   Requirements Analysis (Stage 2)

Based on the established goals, it is required to define a set of requirements (step 2.1) that can be quantitative and qualitative that allow analysts to evaluate them throw metrics. Requirements can be of three types: performance, flexibility, and agility (described in Sect. 1).

Furthermore, the analysts could identify a subset of hard requirements (step 2.2). In this work, hard requirements are called constraints. The GDB must satisfy these constraints to become a candidate in the pre-selection stage. Some examples of constraints are to support: declarative query languages, specific programming languages, and particular graph algorithms.

Besides, the analysts must determine the metrics to measure each one of the requirements previously defined (step 2.3). Some metrics examples are load data efficiently, storage efficiently, and natively graph algorithms execution.

In order to assign a score to each GDB, we suggest a scoring function. The analyst must assign a weight for each metric (each metric is related to one requirement). The score is obtained by summarizing the product of the weights and the normalized metrics results, as it is shown in Eq. 1.

$$S_d^c(M, W) = \sum_{i=1}^{n} w_i ||m_i||, \tag{1}$$

where S_d^c means the score of evaluating a candidate GDB denoted by c for a given dataset d. The function has as input a list (M) of evaluation results of each metric ($m_i \in M$), and a list of weights (W) that determine the importance of each metric.

Besides, the weights have the following property $\sum_{i=1}^{n} w_i = 1$ where $w_i \in W$, $n = |W|$ and each w_i is a value between 0 to 1.

$$\|m_i\| = \begin{cases} \dfrac{m_i - \min(M)}{\max(M) - \min(M)} & maximize\ m_i \\ 1 - \dfrac{m_i - \min(M)}{\max(M) - \min(M)} & minimize\ m_i \end{cases} \qquad (2)$$

In this work, we suggest using the min-max normalization or the complement of this normalization in the case of minimizing the use of some resource as it is described in Eq. 2. This normalization function results in 0 if it is the lowest score and 1 to the highest score. This function will be used in the GDB selection stage (stage 5). Furthermore, the score function and the normalization function can be customized according to the problem to solve.

3.3 GDB Pre-selection (Stage 3)

In this stage, the inputs are a set of constraints defined in the previous stage (stage 2) and a GBD catalog. The catalog consists of all those databases that analyst believe can solve the problem defined in stage 1. Only the GDBs that accomplish all the constraints will be part of the pre-selected GDBs which are going to be considered for the benchmark. The pre-selected GDB catalog will be the output of this stage and input for the next one.

3.4 Benchmarking (Stage 4)

This stage has the pre-selected GDB catalog and the metric definition for each requirement as inputs. The purpose of the stage is to search or build datasets useful to evaluate the GDBs with the metrics established in step 2.3. Moreover, in this stage, the environment of the benchmark has to be defined and configured, such as operating system, RAM, GDB configuration, create indexes on data, among others.

On the other hand, to guarantee a fair comparison between GDBs, it is required to design and implement a set of tests that make possible to use the same set of data for all pre-selected GDBs; for example, a loading performance test requires that all the GDBs load the same amount of data in all tests.

To automate the evaluation of the metrics for each GDB, we suggest the use of some scripts that execute the benchmarks and gather the information of each metric for all the scenarios.

The output of this stage is a table per scenario with the results of the evaluation for each metric for all GDB.

3.5 GDB Selection (Stage 5)

In this stage, the inputs are the results of the evaluations performed over the GDBs of the previous stage, and the scoring function defined in stage 2. Once the scoring function has been applied to all GDBs with each dataset, the final score per GDB is computed by averaging the result of all scenarios (datasets). Hence, it is obtaining a single value per GDB, which is stored as an ordered list of scores and sorted from maximum to minimum. The highest score indicates that the GDB accomplish better the established requirements. Finally, the analysis has all the information to choose one GDB that will be used to solve the problem defined in stage 1.

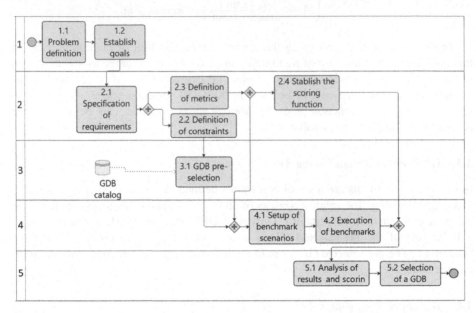

Fig. 1. The process to select the GDB that suits best the business requirements. It is composed of five stages: Problem analysis (1), Requirement analysis (2), GDB Pre-selection (3), Benchmarking (4), and GDB selection (5).

4   Case Study and Results

To show how the selection process of GDB can be applied in a specific case, each stage of the process is described step by step until we can identify the most appropriate GDB to solve a defined problem.

In stage 1, the problem has to be specified by answering the question of why the GDB is needed? In our case study, the problem consists of analyzing a social network to identify the weakly connected components, to obtain the most influential users, and the possibility to explore the neighborhood of some given node. In the next step of this stage, the goals need to be declared. The goals extracted from the previously defined problem are: to store data extracted from a social network, identify weakly connected components, identify opinion leaders or critical users, and to explore the neighborhood of the elements in the social network.

Stage 2 starts with the previously defined goals, and the requirements are specified. In this case study, the requirements are related to GDB performance, as listed in Table 1. However, it is possible to include requirements associated with flexibility and agility.

Table 1. List of performance requirements with its corresponding metrics and weights used in the case study.

Requirements of performance	Metric	Weights
Large-scale data loading	Loading time (min.)	25%
Use of disk storage	Disk usage (GB)	20%
Obtain the neighborhood in 1 hop	Execution time (sec.)	5%
Obtain the neighborhood in 2 hops	Execution time (sec.)	5%
Obtain the neighborhood in 3 hops	Execution time (sec.)	5%
Execution of weakly connected components	Execution time (sec.)	20%
Execution of PageRank algorithm	Execution time (sec.)	20%

Furthermore, in this stage, define the constraints to pre-select the GDB catalog. In this case, the restrictions for GDB are to be an open-source or free, provide a declarative query language, support Java as a programming language, and compute graph algorithms like neighborhood to k-hops, PageRank, and weakly connected components.

Another step (2.3) in the same stage is the definition of the metrics. For this example, we define the metrics in terms of execution time and the required storage space, as shown in Table 1.

Finally, in this stage, the score function is established. In this case study, we use the function suggested in Sect. 3, and the associated weight for each requirement is shown in Table 1.

Table 2. Results of the constraints evaluation on the initial GDB catalog. The first three GDBs accomplish all the constraints. Therefore, these GDBs are pre-selected for the next stages of the selection process.

Graph database	Graph query language	Native graphs triplets	Graph algorithms	Supports Java	Open-source
JanusGraph	X	X	X	X	X
Neo4j	X	X	X	X	X
TigerGraph	X	X	X	X	X
GraphDB	X	X	–	X	X
Mark Logic	X	X	–	X	X
SQL server	–	–	–	X	X

In stage 3, for demonstration purposes, we chose six accessible databases that support graphs. The databases are JanusGraph, Neo4j, TigerGrpah, GraphDB, Mark-Logic, and Microsoft SQL. All databases have to satisfy the constraints of the previous stage. In Table 2, it can be observed that only JanusGraph, Neo4j, and TigerGraph accomplish all constraints, and these can be considered as the pre-select GDBs to be evaluated in the next stage.

In stage 4, the scenarios for evaluation must be defined and configured. For this study, the benchmark setup has been based on the TigerGraph project [9]. The benchmark is conformed of three main steps: the setup, the data loading tests, and the query performance tests.

In this work, we use two datasets published by TigerGraph [9]. The first is a real dataset of the social network Twitter, and the second is a synthetic dataset. The used datasets in this work are described in Table 3.

Table 3. Datasets description used to benchmark pre-selected GDBs.

Dataset	Description	Nodes	Edges	Raw size (MB)
Twitter	A directed graph of social network users	41.6 M	1.47 B	24375
Graph500	Synthetic Kronecker graph	2.4 M	67 M	967

Table 4. Benchmarks results using the Twitter dataset. The last three columns are the outcomes of metric evaluation on pre-selected GBDs. The last row shows the score computed for each GDB.

Metric	JanusGraph	Neo4j	TigerGraph
Loading time (min.)	238.400	74.000	40.300
Disk usage (GB)	33.000	30.000	9.500
K-Neighborhood (sec.)	1.420	0.377	0.017
Two-Hop K-Neighborhood (sec.)	40398.000	5.260	0.330
Three-Hop K-Neighborhood (sec.)	1600.600	99.700	2.900
Weakly connected components (sec.)	86400.000	1545.100	47.900
Page rank query (sec.)	86400.000	614.900	166.030
Score (0–1)	0.000	0.763	1.000

The Twitter dataset has over 41 million nodes and almost 1.5 billion edges; the dataset represents real user connections taken from the social network while Graph500 is a synthetic graph dataset which has 2.4 million nodes and 67 million connections across the data.

Table 5. Benchmarks results using the Graph500 dataset. The last three columns are the outcomes of metric evaluation on pre-selected GBDs. The last row shows the score computed for each GDB.

Metric	JanusGraph	Neo4j	TigerGraph
Loading time (min.)	19.200	6.000	3.700
Disk usage (GB)	2.500	2.300	0.500
K-Neighborhood (sec.)	0.195	0.018	0.003
Two-Hop K-Neighborhood (sec.)	13.950	4.400	0.066
Three-Hop K-Neighborhood (sec.)	1965.500	58.500	0.405
Weakly connected components (sec.)	1491.400	65.500	2.880
Page rank query (sec.)	2279.051	31.200	12.800
Score (0–1)	0.000	0.752	1.000

In both cases, the graph's raw data is formatted as a single tab-separated edge list, and no properties or attributes are attached to edges.

In order to obtain a reliable result, it is required to perform a fair comparison between GDBs. The first condition in order to satisfy such equity is to install the systems on the same hardware and software conditions.

Once the benchmark is set up, it is essential to select an environment which satisfies minimum hardware and software requirements for each database, and this can be checked on each database official requirements publication. Our experiments run on Amazon EC2 instances; each system uses the same number of CPUs, memory, and network bandwidth. The selected operating system must support a native installation of all pre-selected GDBs. In this case, we use Ubuntu 18.04 LTS. Once the databases are configured, the scripts are executed. The results of the benchmark are shown in Table 4 for the first scenario (Twitter), and Table 5 for the second scenario (Graph500). In the case of the Twitter dataset, for JanusGraph, we stopped the execution of the algorithms Page Rank and weakly connected components after 24 h.

Finally, in stage 5, the score function suggested in Sect. 3 is applied to the normalized values, which are the results from the previous stage. Therefore, we obtain, for each scenario, a score per GDB as can be seen in the last row of Tables 4 and 5.

The resulting scores of both tables are averaged, and thus giving a final score that evaluates the GDB in the general point of view. In our case study, the final scores of the pre-selected GDBs are JanusGraph 0.000, Neo4j 0.758, and TigerGraph 1.000. Therefore, according to these results, TigerGraph accomplishes better the requirements. Thus it is selected to solve the defined problem.

5 Conclusions

While getting information about the most popular graph database engines, we have realized that it is common that new products appear and try to get a piece of this evolving market. The proposed selection process shows that interested people in graph database systems can get a quick idea of the strengths and limitations that each database exposes. Our work can help analysts to understand these pros and cons and help them in the process of using an adequate product based on their needs.

The proposed selection process consists of the following stages: problem definition, requirements analysis, GDB pre-selection, benchmarking, and finally, the GDB selection.

Our novel selection process is a useful guide for who needs to evaluate and select a GDB, and it is based on metrics that make possible to ensure that the requirements are fulfilled.

We applied the selection process in a case study of the social network domain, using two datasets to perform the benchmark. Besides, we described how to apply each step of the process until obtaining the final score, which indicates the best GDB for our problem.

The process is flexible enough to be applied to several problems related to a large amount of highly connected data.

As future work, we consider the possibility of implementing a software tool for the (partial) automation of the proposed selection process. The tool can include a set of standard requirements with their corresponding metric. Moreover, we would like to consider the possibility to create building blocks by specifying the input, process, and output. These building blocks may be customized for each GDB and can be used with a friendly user interface where these blocks can be dragged and dropped during the benchmark definition. On the other hand, we would like to test our selection process using FOAF-based dataset in order to benchmark GDB against RDF graph databases.

References

1. Lourenço, J.R., Cabral, B., Carreiro, P., Vieira, M., Bernardino, J.: Choosing the right NoSQL database for the job: a quality attribute evaluation. J. Big Data 2(1), 18 (2015)
2. Baharu, A., Sharma, D.P.: Performance metrics for decision support in big data vs. traditional RDBMS tools & technologies (IJACSA). Int. J. Adv. Comput. Sci. Appl. 7(11), 222–228 (2016)
3. Cattell, R.: Scalable SQL and NoSQL data stores. SIGMOD Rec. 39(4), 12–27 (2011)
4. Han, J., Haihong, E., Le, G., Du, J.: Survey on NoSQL database. In: 6th International Conference on Pervasive Computing and Applications, pp. 363–366. IEEE, Beijing, China (2011)
5. Junghanns, M., Petermann, A., Neumann, M., Rahm, E.: Management and analysis of big graph data: current systems and open challenges. In: Handbook of Big Data Technologies. Springer, Sydney (2017)
6. Guia, J., Soares, V., Bernardino, J.: Graph databases: Neo4j analysis. In: Proceedings of the 19th International Conference on Enterprise Information Systems, pp. 351–356 (2017)
7. Hayes, J., Gutierrez, C.: Bipartite graphs as intermediate model for RDF. In: International Semantic Web Conference, pp. 47–61. Springer, Heidelberg (2004)
8. Angeles, R., Gutierrez, C.: Querying RDF data from a graph database perspective. In: European Semantic Web Conference, pp. 346–360. Springer, Heidelberg (2005)
9. Vicknair, C., Macias, M., Zhao, Z., Nan, X., Chen, Y., Wilkins, D.A.: Comparison of a graph database and a relational database: a data provenance perspective. In: Proceedings of the 48th annual Southeast regional conference, p. 42. ACM, Mississippi (2010)
10. Batra, S., Tyagi, C.: Comparative analysis of relational and graph databases. Int. J. Soft Comput. Eng. 2, 509–512 (2012)
11. Nayak, A., Poriya, A., Poojary, D.: Type of NoSQL databases and its comparison with relational databases. Int. J. Appl. Inf. Syst. 5, 16–19 (2013)
12. Fernandes, D., Bernardino, J.: Graph databases comparison: AllegroGraph, ArangoDB, InfiniteGraph, Neo4 J, and OrientDB. In: Proceedings of the 7th International Conference on Data Science, Technology and Applications, pp. 373–380, Porto, Portugal (2018)
13. Neo4j vs. Dgraph – The Numbers Speak for Themselves. https://blog.dgraph.io/post/benchmarking-neo4j/
14. Benchmarking Graph Analytic Systems: TigerGraph, Neo4j, Neptune, JanusGraph, and ArangoDB. https://www.tigergraph.com/benchmark/
15. Maxville, V., Armarego, J., Lam, C.P.: Applying a reusable framework for software selection. IET Softw. 3(5), 369–380 (2009)
16. Lee, Y.-C., Tang, N.-H., Sugumaran, V.: Open source CRM software selection using the analytic hierarchy process. Inf. Syst. Manag. 31(1), 2–20 (2014)

Influence of Belbin's Roles in the Quality of the Software Requirements Specification Development by Student Teams

Raúl A. Aguilar Vera[1]([⊠]), Julio C. Díaz Mendoza[1],
Mirna A. Muñoz-Mata[2], and Juan P. Ucán Pech[1]

[1] Facultad de Matemáticas, Universidad Autónoma de Yucatán,
97000 Mérida, Yucatán, Mexico
{avera, julio.diaz}@correo.uady.mx,
juan.ucan@correo.uady
[2] Centro de Investigación en Matemáticas, Unidad Zacatecas,
Av. Universidad 222, 98068 Zacatecas, Mexico
mirna.munoz@cimat.mx

Abstract. The article presents a controlled experiment in which the convenience of using the Belbin Role Theory for the integration of work teams, in the tasks of software requirements process, is explored. The study is developed in an academic environment with students of the Software Engineering degree and analyzes the differences between the quality of the software requirements specification document generated by integrated teams with compatible roles (Belbin Teams: BT, according to the Belbin Theory) with the Traditional Teams (TT), in our case, teams integrated with students selected at random. The results of the study provide evidence that the quality of the ERS documents generated by the Belbin Teams is significantly better than that generated by the traditional equipment.

Keywords: Belbin´s roles · Controlled experiment · Improvement process software · Requirement process · Software Engineering

1 Introduction

After half a century of its conception, Software Engineering (SE) has managed to accumulate a body of knowledge accepted by professionals and researchers of the discipline, which integrates a set of areas linked to development processes (Requirements, Design, Construction, Testing and Maintenance) and management processes (Quality, Configuration, Software Process) associated to the aforementioned areas [1, 2]. Nowadays, there is a constant search for new methods, techniques, tools and good practices of SE that allow a continuous improvement both the processes and the products that are derived from them.

The improvement of the software process has been studied considering several factors; one that has had a singular interest in recent years is related to the influence of the human factor in the software development process. This work is derived precisely from our interest in deepening in this research area, particularly we are interested in

© Springer Nature Switzerland AG 2020
J. Mejia et al. (Eds.): CIMPS 2019, AISC 1071, pp. 91–101, 2020.
https://doi.org/10.1007/978-3-030-33547-2_8

exploring the use of Belbin's theory for the integration of work teams, and analyze its possible influence on those tasks related to both processes of software development as in software management.

The remaining part of this document is organized as follows, Sect. 2 presents a general description of related work; the Sect. 3 briefly describes the software requirements process. Sections 4 and 5 present the planning and execution of the controlled experiment, respectively. Section 6 presents the statistical analysis of the data obtained of the controlled experiment, as well as the results thereof. Finally, Sect. 7 the conclusions of the study are presented and then the references are listed.

2   Related Work

Some researchers [3–5] claim to have identified roles that describe the behavior of individuals in work teams —team roles— and although there is no evidence that they are particularly associated with any type of activity, their absence or presence is says that it has significant influence on the work and the achievements of the team [6]. Among the proposals on team roles, Belbin's work [5, 7] is surely the most used among consultants and trainers, its popularity is that, it not only offers a categorization of roles (Sharper: I, Implementer: IM, Completer-Finisher: CF, Plant: P, Monitor-Evaluator: ME, Specialist: SP, Chairman: CH, Resource Investigator: RI & Teamworker: TW) but also describes a series of recommendations for the integration of work teams, which are known as the Belbin Role Theory.

Pollock in [8] explores whether diversity of Belbin's roles and personalities among the members of a team of information systems students can improve the effectiveness of the team. He concludes that diversity does not have a significant influence on the effectiveness of the team; however, he says that the presence of certain roles, such as the Sharper, Chairman and Completer-Finisher, can increase effectiveness.

Aguilar in [9] reports that the collaboration skills presented by groups formed on the basis of role theory (Belbin Role Theory), is significantly greater than that presented by the groups formed solely taking into account functional roles; reports that groups formed on the basis of Belbin Role Theory spend more time than groups formed with functional roles in their work sessions.

Estrada and Peña [10] conducted a controlled experiment with students, in which they observed the performance of team roles in tasks related to the software development process, as part of their conclusions reported that some roles have greater contribution with certain activities, as was the case of the Implementer role in the coding task.

In a previous work [11] we compared, in an academic environment, the quality of the readability of the code generated by teams integrated with the Belbin Theory (EB), in contrast to randomly formed teams (ET); among the results obtained, it was reported that EB teams show significantly better results than ET teams. In another work, we explore with the task of measuring software [12]. We reported two controlled experiments performed with the same task, measuring software using the function point technique, however, we did not find conclusive information on the inflow of Belbin roles in the measurement task, apparently the way to integrate the groups, has no influence on the correct measurement of the software.

3   The Software Requirements Process

The purpose of the software requirements process is to define the set of functionalities and restrictions that address the definition of the software product [2]. The final product of the requirements process is known as the Software Requirements Specification, and the IEEE Std 830-1998 establishes the content and qualities of a good software requirements specification (ERS) document [13].

The software requirements development process consists of four phases [14]. The first stage of the process, called Elicitation of requirements, is concern with the identification of the real needs and operating restrictions expressed by the stakeholders of an organization. To obtain this information, the Software Engineer designs a strategy in which he combines a set of traditional, contextual, group, and/or cognitive techniques [15]. The second stage of the process, Analysis, aims to ensure the quality of the requirements before incorporating them into the specification document. In this stage, the limits of the software system and its interaction with the environment are defined. User requests are translated into software requirements. The Specification stage consists on documenting the agreed software requirements, at an appropriate level of detail, and a document model written in terms understandable to stakeholders is usually used (e.g. IEEE Std 830-1998). Finally, in the last phase of the process, Validation, a careful review is made of a set of quality attributes linked both to the individual requirements and to the document in an integral manner; the objective is to identify problems in the software requirements specification document before it is used as a basis for the design of the System.

Quality is difficult to define, it is subjective and depends on the different viewpoints and the context on the construct to which the concept is applied [16]. In the case of the product generated with the requirements process, we can use the IEEE 830-1998 standard as a starting point to define the quality of the requirements (see Table 1).

Table 1. Examples of quality attributes for ERS

Quality attributes	Definition
Completeness	All important elements that are relevant to fulfill the different user's tasks should be considered. This includes relevant functional and non-functional requirements
Consistency	The stated requirements should be consistent with all other requirements, and other important constraints
Correctness	The requirements that are implemented have to reflect the expected behavior of the users and customers
Traceable	It should be possible to refer to the requirement in an easy way, both for artifacts derived from the requirement, and from which it is derived
Unambiguity	The requirements should only have one possible interpretation
Verifiability	There must be a process for a machine or a human being to check whether the requirement is met or not

The verification process carried out by the software engineer must be done in two levels. At a first level, quality attributes must be evaluated individually for software requirements (e.g. Traceable, Unambiguity, Verifiability) and at the second level, quality attributes are verified to the requirements specifications document in a comprehensive manner (e.g. Completeness, Consistency, Correctness).

4 Planning of the Experiment

A controlled experiment was designed with the purpose of explore the possible influence of the use of the Belbin Theory for the integration of software development teams, particularly, in the evaluation of the product of the requirements phase, that is, the quality of the ERS document.

4.1 Objective, Hypothesis and Variables

With the aim of exploring whether the integrated work teams based on the Belbin Role Theory —which we will call Compatibles— generate ERS documents significantly different in terms of their quality from those obtained by the integrated teams without using any particular criteria —which we will call Traditional— a pairs of hypotheses was proposed:

H_0: The quality of ERS document development by the Traditional Teams is equal to the quality of ERS document development by the Belbin Teams.
H_1: The quality of ERS document development by the Traditional Teams differs from to the quality of ERS document development by the Belbin Teams.

The factor controlled by the researchers in this experiment is the integration mechanism of the software development teams, which has two alternatives: (1) Belbin Teams, and (2) Traditional Teams. On the other hand, the response variable, Quality of the ERS document, was obtained through the validation of each of the products generated by the participating teams.

4.2 Checklist for Evaluation of ERS Quality

In order to evaluate the products generated by the development teams, a checklist was designed with a set of items that evaluate various quality attributes that are considered necessary or appropriate for the ERS. The checklist uses a Likert scale to collect the opinions (evaluations) of the reviewers of the ERS.

The instrument is used to verify both quality attributes linked to the requirements (16 items) and linked to the specification document (8 items). Additionally, three items are included with which the correspondence between the use cases and the software requirements is verified.

4.3   Experimental Design

A Factorial Design with "one variation at a time" was used, the independent variable corresponds to the way the work teams were integrated. In the study there are two experimental treatment: BT (Belbin Teams) and TT (Traditional Teams). Table 2 presents the experimental design that we will use.

Table 2. Experimental design

Treatment	Teams
Belbin Teams (BT)	BT1, BT2, BT3, ...
Traditional Teams (TT)	TT1, TT2, TT3, ...

5   Execution of the Experiment

According to the plan of the experiment, we worked with a group of students of a Bachelor in Software Engineering, enrolled in the on the topics of the subject Development of Software Requirements. Throughout the course, one of the researchers teaches to the group the topics related to the software requirements development process, and from the third unit —of the program of the subject— the students, in work teams formed by the teacher, began to apply the requirements process to a real-world problem selected by the students, in agreement with the teacher. The final product of the assignment consisted of the ERS document based on the IEEE Std 830-1998, and includes the Use Cases according to the services specified in the ERS.

5.1   Participants/Subjects

The participants in the experiment were thirty-four students enrolled in the subject Development of Software Requirements in the August-December 2018 semester, subject located in the fifth semester of the curricular map.

With the students enrolled in the course, eight software development teams of four members each were formed; we used the information obtained —primary roles in students— after the administration of the self-perception inventory of Belbin (instrument proposed by Belbin to identify the role that the subject that answers it in a work team can play) and we integrate four teams with compatible roles, that is, roles that have no conflict between them and that according to Belbin, as a team they can take advantage of their individual characteristics (Belbin Teams: BT) and four additional teams with students assigned in a random way (Traditional Teams: TT). Two students integrated the ninth team of the group, but they would not participate in the study. For personal reasons, one of the students who belonged to one of the BT, left the subject when they were working on the first phase of the requirements process, and it was necessary to make an adjustment in the team; team III was discarded for the study.

Since the measurements would be obtained on the products generated by the development teams, the experimental subjects in this case were the seven work teams integrated by the researchers. The conformation of the three integrated teams based on the Belbin Role Theory is illustrated in Fig. 1; letters inside the circles indicate the identifier used for the team roles played by each participant (see the first paragraph of the related works section), and thus the conformation of the compatible team (experimental subjects) can be observed. The number between the four circles is an identifier for the team.

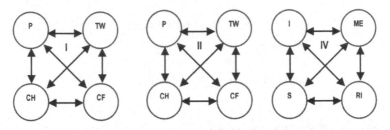

Fig. 1. Integrated teams with Belbin Theory (Belbin Teams)

Using the selected experimental design as a reference, Table 3 illustrates the assignment of the seven student teams to each of the treatments.

Table 3. Experimental subjects by treatment.

Treatment	Teams
Belbin Teams (BT)	I, II, IV,
Traditional Teams (TT)	V, VI, VII, VIII

5.2   Execution of the Study

The study was conducted at four different moments; in a first session, the self-perception instrument was administered to identify the primary team roles for the participants; in a second stage, the teacher worked with the students throughout the course on the topics of the subject, and the seven teams developed the activities of the requirements process. In a third moment, the teams delivered the ERS document according to the agreement with the professor. Finally, two independent reviewers to the researchers used the checklist to evaluate the quality of the products.

To have an objective evaluation, two analysts were invited to evaluate the ERS documents using the checklist designed by the team of researchers, based on a set of quality factors commonly used in the requirements phase (see Table 1). The results of the evaluations carried out independently were averaged to generate a metric as a response variable (see Table 4).

Table 4. Quality metrics of the ERS documents generated by the development teams

Team	Validation 1	Validation 2	Quality ERS
I	106	107	106.5
II	119	126	122.5
IV	121	106	113.5
V	99	95	97
VI	108	89	98.5
VII	117	105	106
VIII	91	89	90

6 Statistical Analysis and Results

This section presents both the descriptive statistical analysis of the measurements collected and the inferential statistical analysis.

Table 5 presents some of the most important measures of central tendency and variability; we should note that the Mean BT is greater than that of the TT, however, the former has greater Variance, although the Range of values is the same. We can also observe that the Median presents a value very close to the Mean.

Table 5. Summary statistics

Factor	Count	Mean	Median	Variance	Minimum	Maximum	Range
BT	3	114.167	113.5	64.3333	106.5	122.5	16.0
TT	4	97.875	97.75	43.0625	90.0	106.0	16.0

In order to visually compare the two treatments, we generated a boxplot, this chart allows us to observe the dispersion and symmetry of both data sets (see Fig. 2).

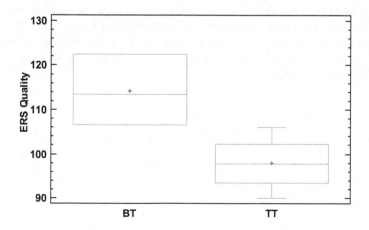

Fig. 2. Boxplot for development teams

In Fig. 2 we can see that do not present overlaps in the treatments, so it is very likely that they present differences between them. We can also observe that no large differences were observed in the dispersion of the treatments; On the other hand, both treatments present symmetry.

In order to evaluate the differences observed in the dependent variable —as a product metric— with the descriptive analysis and determine if they are significant from the statistical perspective, the following statistical hypotheses were raised:

$$H0: \mu_{BT} = \mu_{TT}; \quad H1: \mu_{BT} <> \mu_{TT}$$

We proceeded to use the one-way ANOVA. Table 6 present the results of these analyzes.

Table 6. ANOVA for quality by teams.

Source	SS	Df	MS	F	P-value
Between groups	455.003	1	455.003	8.82	0.0311
Within groups	257.854	5	51.5708		
Total (Corr.)	712.857	6			

The ANOVA table decomposes the variance of the variable under study into two components: an inter-group component and an in-group component. Since the P-value of the F-test is less than 0.05 (p-value = 0.0311), we can reject the null hypothesis and affirm in both cases that there is a statistically significant difference between the mean of the variable under study between a level of Treatment and other, with a 5% level of significance.

The ANOVA Model has associated three assumptions that it is necessary to validate before using the information it offers us; the assumptions of the model are: (1) The experimental errors of your data are normally distributed, (2) Equal variances between treatments (Homoscedasticity) and (3) Independence of samples.

To validate the first assumption, we will use the normal probability graph. It is a graphical technique for assessing whether or not a data set is approximately normally distributed. As can be seen in the graph of Fig. 3, the points, in both graphs, do not show deviations from the diagonal, so it is possible to assume that the residuals have a normal distribution in both cases.

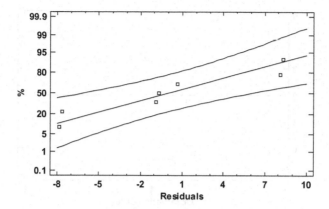

Fig. 3. Normal probability plot

In the case of Homoscedasticity, we generate a Residuals vs. Fitted plot, and we observed if it is possible to detect that the size of the residuals increases or decreases systematically as it increases the predicted values. As we can see in the graph of Fig. 4, no pattern is observed by which we can accept that the constant variance hypothesis of residuals is met.

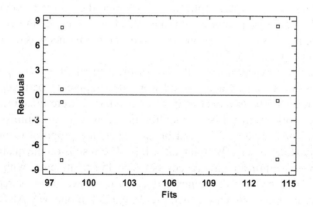

Fig. 4. Residuals vs. Fitted plot

Finally, to validate the assumption of data independence, we generate a residuals vs. Order of the Data plot. In this case, we observe if it is possible to detect any tendency to have gusts with positive and negative residuals; in the case of our analysis, we can see in Fig. 5 that no trend is identified, so it is possible to assume that the data come from independent populations.

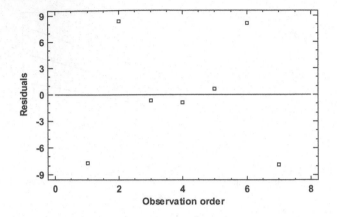

Fig. 5. Residuals vs. Order of the Data plot.

With the three valid assumptions for the information obtained in the experiment, it is possible to use the information generated with the model for our conclusions.

7   Conclusions

The requirements process seeks to satisfy the needs of the stakeholders by specifying the functionalities of a software application, for which the Software Engineer interacts through the elicitation techniques with the stakeholders in an individual or group manner, and the process of analysis and specification requirements is done with various iterations between peer reviewers.

In this paper, we present a controlled experiment in which we compare the quality of work teams in tasks related to software development, particularly, software requirements process. The two treatments to be compared, were linked with the way of integrating the work teams; firstly, the traditional way of randomly assigning its members (Traditional Teams: TT), and on the other, the proposal to integrate teams using the Belbin Role Theory (Belbin teams: BT). The results on the quality of the ERS document showed significant differences between the treatments, with better results from the BT. To carry out the controlled experiment we used paired comparison designs, and for information analysis, we use the Model of one way ANOVA, which is used to compare two means from two independent (unrelated) groups using the F-distribution. The obtained in this experiment coincide with those observed in [11], but contrast with what was reported in [12]. Possibly the task of measuring software, developed in teams, does not require the development of high levels of interaction among team members; however, the coding process requires a certain level of inter-action among the programmers, but even more, the process of software requirements is an Activity with a strong social component.

It is important to mention that when we use experimentation as an empirical methodology, two of the circumstances we encounter are those related to: (1) the non-large sample size, (2) the need for convenience sampling; although, the ideal in statistics to have a representative sample of the population would be: (1) using an appropriate type of sampling and (2) selecting a representative sample size, it is practically impossible to perform in our field. By virtue of the problems described above, the strategy adopted for some decades, is the development of experimental replicas, its purpose is to verify or refute the findings previously observed [17] and thereby generate knowledge about the phenomenon; on the other hand, developing a series of experimental replicas allows us to refine research hypotheses and thereby obtain more specific knowledge about the phenomenon.

As part of the work in progress, the research group has begun to carry out experimental replicas to refine the results obtained in the exploratory studies [12].

References

1. Abran, A., Moore, J.W., Bourque, P., Dupuis, R., Tripp, L. (eds.): Guide to the software engineering body of knowledge – 2004 Version. IEEE Computer Society (2004)
2. Bourque, P., Fairley, R. (eds.): Guide to the Software Engineering Body of Knowledge (SWEBOK V3.0). IEEE Computer Society, Washington, D.C. (2014)
3. Mumma, F.S.: Team-work & team-roles: what makes your team tick? King of Prusia. HRDQ, PA (2005)
4. Margerison, C.J., McCann, D.J.: Team management profiles: their use in managerial development. J. Manag. Dev. **4**(2), 34–37 (1985)
5. Belbin, M.: Management Teams. Wiley, New York (1981)
6. Senior, B.: Team roles and Team performance: is there 'really' a link? J. Occup. Organ. Psychol. **70**, 85–94 (1997)
7. Belbin, M.: Team roles at Work. Elsevier Butterworth Heinemann, Oxford (1993)
8. Pollock, M.: Investigating the relationship between team role diversity and team performance in information systems teams. J. Inf. Technol. Manag. **20**(1), 42–55 (2009)
9. Aguilar, R.: Una Estrategia Asistida por Entornos Virtuales Inteligentes. Doctoral thesis. Polytechnic University of Madrid (2008)
10. Estrada, E., Peña, A.: Influencia de los roles de equipo en las actividades del desarrollador de software. Revista Electrónica de Computación, Informática, Biomédica y Electrónica **2**(1), 1–19 (2013)
11. Aguileta, A., Ucán, J., Aguilar, R.: Explorando la influencia de los roles de Belbin en la calidad del código generado por estudiantes en un curso de ingeniería de software. Revista Educación en Ingeniería **12**(23), 93–100 (2017)
12. Aguilar, R., Díaz, J., Ucán, J.: Influencia de la Teoría de Roles de Belbin en la Medición de Software: Un estudio exploratorio. Revista Ibérica de Sistemas y Tecnologías de la Información **31**, 50–65 (2019)
13. IEEE Computer Society Software Engineering Standards Committee: IEEE Recommended Practice for Software Requirements Specifications. IEEE Std 830-1998 (1998)
14. Aguilar, R., Oktaba, H., Juarez, R., Aguilar, J., Férnandez, C., Rodriguez, O., Ucán, J.: Ingeniería de Software. In: En Pineda, L. (ed.) La computación en México por especialidades académicas. Academia Mexicana de Computación, Capítulo V (2017)
15. Yousuf, M., Asger, M.: Comparison of various requirements elicitation techniques. Int. J. Comput. Appl. **116**(4) (2015)
16. Kitchenham, B., Pfleeger, S.: Software quality: the elusive target. IEEE Softw. **13**(1), 12–21 (1996)
17. Gómez, O., Juristo, N., Vegas, S.: Understanding replication of experiments in software engineering: A classification. Inf. Softw. Technol. **56**(8), 1033–1048 (2014)

A Comparative Review on the Agile Tenets in the IT Service Management and the Software Engineering Domains

Manuel Mora[1](✉), Fen Wang[2], Jorge Marx Gómez[3],
and Oswaldo Díaz[4]

[1] Autonomous University of Aguascalientes, Information Systems,
Av. Universidad 940, 20131 Aguascalientes, Mexico
dr.manuel.mora.uaa@gmail.com
[2] Central Washington University, Technology and Administrative Management,
400 E. University Way, Ellensburg, WA 98926, USA
Fen.Wang@cwu.edu
[3] University of Oldenburg, Informatics, Ammerländer Heerstr. 114–118,
26129 Oldenburg, Germany
jorge.marx.gomez@uol.de
[4] INEGI, Data Center Central, Av. Héroe de Nacozari Sur #2301,
20276 Aguascalientes, Mexico
oswaldo.diaz@inegi.org.mx

Abstract. A rigor-oriented paradigm, in the Information Technology Service Management (ITSM) domain, has permeated in multiple international organizations. In contrast, in the Software Engineering (SwE) domain, the agile paradigm has complemented or replaced the rigor-oriented one in the last two decades. Recently shortened ITSM methods (i.e. FitSM and IT4IT) have emerged. However, due to their novelty, there is a scarce analysis of literature on their main assumed agile tenets. In this study, we identified and compared such tenets (i.e. aims, values, and principles) of the agile SwE and ITSM paradigms as well as the adherence to them from a representative literature on agile SwE and ITSM methods. Our results identified a high and low adherence respectively in the SwE and ITSM methods. Thus, a call for a robust theoretical foundation on agile tenets like those found in the SwE methods is required for the ITSM methods.

Keywords: Agile paradigm tenets · Agile ITSM methods · Agile software engineering methods · FitSM · IT4IT · Representative literature analysis

1 Introduction

ITSM has been defined as "the implementation and management of quality IT services that meet the needs of the business" [1; p. 16]. IT services are services delivered through IT, process and people. Services, in the domain of ITSM, are "means of delivering value to customers by facilitating outcomes customers want to achieve without the ownership of specific costs and risks" [1; p.13]. Value is realized when the

© Springer Nature Switzerland AG 2020
J. Mejia et al. (Eds.): CIMPS 2019, AISC 1071, pp. 102–115, 2020.
https://doi.org/10.1007/978-3-030-33547-2_9

IT service is appreciated with the adequate levels of utility (fit for purpose) and warranty (fit for use). The utility of an IT service refers to what the service does and warranty to how it is delivered [1]. The utility of an IT service is manifested by a user-task performance improvement and/or a user-task reduction of constraints. Warranty of an IT service is manifested by its expected availability, capacity, and reliability - continuity and security- levels. Thus, an IT service produces value when both expected utility and warranty are manifested. The occurrence of a single core attribute either the utility or warranty alone cannot produce value to IT users.

There are several ITSM methods potentially for being implemented in the organizations. The main ones used in the last 15 years by thousands of international organizations [2, 3] are ITIL v2, ITIL v3-2011 and ISO/IEC 20000. Whereas they present differences, all of them share a common rigor-oriented and process-heavy paradigm. Thus, despite the organizational benefits reported by its utilization such as IT service quality improvement, IT service management cost reduction, and IT service user satisfaction increment, [3], their implementation demands significant organizational resources (economic, human and technological ones) and efforts (large implementation time periods), which presents great challenges to small and medium-sized organizations for implementing them successfully [4].

In contrast, in the SwE domain, the rigor-oriented and heavy-process paradigm has been strongly replaced by the Agile one [5]. Scrum and XP are the two most used methods [5]. A core study on the Agile paradigm has suggested a shared aim, values set, principles set, and practices set, leading toward a theoretical mature field [6]. Furthermore, additional research has been conducted to identify the project's profiles suitable for using the Agile paradigm, while for another kind of projects, this Agile paradigm is not recommended [7].

In the ITSM domain, FitSM [8] and IT4IT [9] are two shortened ITSM methods recently appeared in the last 5 years which can be assumed as agile ones. Hence, due to the novelty of these ITSM methods, there is scarce analysis literature on their main tenets regarding to the Agile paradigm [10]. Thus, in this study, we identified and compared the tenets of the agile SwE and ITSM paradigms, as well as the adherence to them from a representative literature agile SwE and ITSM methods. The remainder of this article continues as follows: in Sect. 2, the research goal, questions, and methodology are presented; in Sect. 3, the conceptual results are reported; and in Sect. 4, discussion of results and conclusions are presented.

2 Research Goal, Questions and Methodology

In this study, we used a Representative Literature Review method [11]. This research method differs from a Systematic Literature Review, in its qualitative and interpretative human-based analysis instead of an intensive-quantitative and more automatic-oriented analysis normally conducted by the latter one. This Representative Literature Review method also relies on a selective small sample of very high-quality documents (i.e. highly cited) versus an exhaustive search of documents selected by a Systematic Literature Review [12]. Specifically, this research method was applied with the following research steps: RS.1) to state the research goal and questions; RS.2) to collect relevant (i.e. documents which address agile tenets) and adequate (i.e. published as journal articles, research book chapters, conference proceeding paper or reputed professional empirical survey) documents for analysis; RS.3) to establish analysis pro-formas, analyze each document in the document set, and to fill them; and RS.4) to review, discuss and agree by the research team the pro-formas and logical derivations from them.

2.1   Research Step 1

The research goal of this study was to identify (i.e. a descriptive purpose) and compare (i.e. a critical purpose) the tenets (i.e. aim, values and principles) of the agile paradigms in the SwE and ITSM domains. We considered these elements as fundamental tenets, whereas the study of their practices and roles is planned for as an extension of this current study in the future. Five specific research questions arose as follows: RQ.1) what are the shared aim, values and principles of the agile SwE domain? R.Q.2) what is the extent of adherence of the agile SwE methods literature to its proffered agile tenets? RQ.3) what are the shared aim, values and principles in the agile ITSM domain? R.Q.4) what is the extent of adherence of the agile ITSM methods literature to its proffered agile tenets? And RQ.5) what do similarities and differences exist between both proffered agile SwE and ITSM tenets and their adherence to them by their agile SwE and ITSM methods literature?

2.2   Research Step 2

For collecting documents in the agile SwE domain, we had access to an already document set collected from GoogleScholar research database in the 2000–2015 period [13]. A subset of 43 documents was selected based on their high and clear relevance by focusing on analyzing agile tenets from comprehensive studies, and their high relevance by the number of received citations (i.e. with at least 30 citations per document OR identified as a document highly relevant for the research goal established in this study). This document data set included 3 documents from the AgileAlliance.org website, which is the central repository from the founders of the agile paradigm, and 2 documents from the Scrum and XP inventors (one of them was published in 1999). This 43-document set was limited to journal articles, research book chapters, conference papers or industrial reports. We complemented this 43-document set with a new search for the 2016-2019 period. By practical concordance, we used the same academic search engine (i.e. GoogleScholar). The query used was "(agile software) AND (engineering OR paradigm OR approach OR process OR methods OR development OR survey OR literature OR "success factors")" in the title. This query produced 11 new located documents, for a final agile SwE 54-document set. We did not apply a forward snowball search for this agile SwE 54-document set to keep the analysis practically manageable. Further, this representative SwE 54-document set was considered of high-quality by the huge number of jointly received citations (i.e. over 14,000 citations).

A search in digital academic databases from scratch was necessary for collecting documents in the agile ITSM domain. We defined the practical and methodological selection criteria as follows: (1) to use well-known and available scientific data sources (Scopus, Springer, and GoogleScholar were used); (2) to fix the search period as between the 2000–2019 years; (3) to specify the document's language as English; (4) to limit the type of document to journal article, research book chapter, conference paper or industrial report (i.e. books were excluded); and (5) to query data sources with the following logical statement: "(agile) AND (ITSM OR ITIL OR "ISO 20000" OR FitSM OR IT4IT) within (Title OR Abstract)". We located for Scopus, Springer and GoogleScholar respectively 21, 19 and 34 documents. The research team filtered the available abstracts based on their potential relevance on addressing agile tenets in the ITSM domain, eliminated duplicated documents, and selected 7, 3 and 2 documents respectively from Scopus, Springer, and GoogleScholar. We applied a forward snowball search on the 12 documents, and we located only 2 new relevant documents from the total of the 47 references (i.e. 19, 19, and 9 citations in Scopus, Springer and GoogleScholar). This resultant set was, thus, of 14 documents. We did not apply a backward snowball search procedure because: (1) some references would be out of the fixed period range; (2) our purpose was not doing a historical tracking of the agile ITSM domain, but a thorough analysis of comprehensive studies on agile ITSM tenets; and (3) the low benefit/effort-cost of undertaking this additional search.

For the case of the emergent shortened ITSM methods (FitSM and IT4IT), because of the number of relevant documents was zero, we considered as necessary to include their official descriptive documents and apply an additional special search in the GoogleScholar engine. We selected 10 official descriptive documents (5 per each emergent ITSM method). The special search in GoogleScholar was established as "(FitSM OR IT4IT) within (Content)", and we located 8 relevant documents on FitSM and IT4IT from 153 ones. Thus, for the agile ITSM domain, we collected a filtered 32-document set.

Thus, the joint final document set for analysis was of 86 documents (i.e. 54 for the SwE domain and 32 for the ITSM domain). Table 1 shows the distribution of the relevant (already filtered and non-duplicated) documents collected by domain, search case and search period. The column labeled as "Total of Documents" includes the newly added and already filtered documents obtained from a forward snowball search procedure applied to the agile ITSM search.

For analyzing the 86-document set, we divided it into two stages of review. The first sub-set was of 20 documents and the second one included 66 documents. In this article, we reported the results for the first stage of 20 documents (i.e. 12 and 8 for the agile SwE and ITSM domains respectively to keep the same proportion from the total of 86-document set). For selecting these 20 documents, we classified the type of studies in the following 4 categories: Seminal) seminal studies on the main agile SwE methods (Scrum, XP) or ITSM methods (FitSM, IT4IT) from their originators; Core) core deep studies on the agile SwE or ITSM methods; Comprehensive) comprehensive on breadth studies on the agile SwE or ITSM methods; and Factors) studies on success and benefits Factors for agile SwE or ITSM methods. Finally, Tables 2 and 3 reports the 12 and 8 selected papers for this first stage of analysis.

2.3   Research Steps 3 and 4

The execution of these steps is reported respectively in Sects. 3 and 4 of this paper.

Table 1. Distribution of the 86-document set on the agile SwE and ITSM domains for the period 2000–2019

Domain/Research database	2000–2005	2006–2010	2011–2015	2016–2019	Total of documents	Total of citations
Agile SwE/GoogleScholar} initial document set 2000–2015	10	10	23	0	43	14,161
Agile SwE/GoogleScholar} new 2016–2019 period search	0	0	0	11	11	173
Agile ITSM/{Scopus, Springer, GoogleScholar} new search	0	2	6	6	14	47
Agile ITSM {FitSM, IT4IT} official repository search	0	0	1	9	10	2
Agile FitSM and IT4IT {GoogleScholar} special search	0	0	3	5	8	2

Table 2. List of the 12 selected documents on the agile SwE domain for the period 2000–2019

Ref	Agile domain	Year	Type	Document title
[17]	SwE	1999	Seminal	Embracing Change With Extreme Programming
[18]	SwE	2004	Core	Extreme programming and agile software development methodologies
[19]	SwE	2005	Core	Agile Software Development Methods: When and Why Do They Work
[20]	SwE	2007	Core	Agile software development: theoretical and practical outlook
[21]	SwE	2007	Comprehensive	Usage and perceptions of agile software development in an industrial context: An exploratory study
[22]	SwE	2011	Comprehensive	Signs of agile trends in global software engineering research: A tertiary study
[23]	SwE	2011	Comprehensive	A study of the Agile software development methods, applicability and implications in industry
[24]	SwE	2011	Factors	User acceptance of agile information systems: A model and empirical test
[25]	SwE	2012	Core	A decade of agile methodologies: Towards explaining agile software development
[26]	SwE	2016	Core	Popular agile methods in software development: Review and analysis
[27]	SwE	2018	Core	The rise and evolution of agile software development
[28]	SwE	2018	Factors	Agile methodologies: organizational adoption motives, tailoring, and performance
[28]	SwE	2018	Factors	Agile methodologies: organizational adoption motives, tailoring, and performance

Table 3. List of the 8 selected documents on the agile ITSM domain for the period 2000–2019

Ref	Agile domain	Year	Type	Document title
[29]	ITSM	2006	Core	ITIL vs. Agile Programming: Is the Agile Programming Discipline Compatible with the ITIL Framework?
[30]	ITSM	2014	Core	Improving PL-Grid Operations Based on FitSM Standard
[31]	ITSM	2015	Core	Enterprise ICT transformation to agile environment
[32]	ITSM	2016	Seminal	FitSM: Part 0: Overview and vocabulary
[33]	ITSM	2016	Comprehensive	Information technology service management models applied to medium and small organizations: A systematic literature review
[34]	ITSM	2016	Seminal	IT4IT™ Agile Scenario Using the Reference Architecture
[35]	ITSM	2017	Seminal	Defining the IT Operating Model
[36]	ITSM	2019	Core	IT Service Management Frameworks Compared–Simplifying Service Portfolio Management

3 Conceptual Results

The 12 and 8 selected documents respectively for the agile SwE and ITSM domains were carefully reviewed to locate evidence statements supporting the core elements pursued in the RQ.1 and RQ.3 questions. These core elements were the shared aim, values and principles in the SwE and ITSM domains. For the agile SwE domain, we used a conceptual scheme derived directly from the official agile SwE concepts reported by the Agilealliance.org [14, 15], which is the organization that groups the main proponents of the agile SwE domain. For the agile ITSM domain, at the best knowledge, despite the existence of some similar organization like itSMF international (https://www.itsmfi.org), none agile ITSM manifest has been elaborated by this ITSM organization. Nevertheless, it was located in the agile ITSM literature an initial proposal for an agile ITSM framework [16] based on the agile SwE manifest [14].

Tables 4 and 5 report respectively the overall results of the analysis conducted for answering the five research questions. These tables assessed a qualitative value of HIGH (symbol ■), LOW (symbol ■) or N.A. (symbol □) level, regarding the existence of evidence statements supporting the aim, values and/or principles. A HIGH level was assessed when at least 2 evidence statements referred directly to an agile tenet. A LOW level when just one evidence statement was located. A N.A. (not available) level when no evidence statement was located in the set of analyzed papers. Tables 6, 7, 8 and 9, reported in the Appendix, show the evidence statements collected from the 20 reviewed documents.

Table 4. Overall assessment of the 12 selected documents on the agile tenets in the agile SwE methods literature

Type	Tenet	Sem.	Core	Comp.	Factors
Aim	*We are uncovering better ways of developing software by doing it and helping others do it*	■	■	■	■
Values	*V1. Individuals and interactions over processes and tools. V2. Working software over comprehensive documentation. V3. Customer collaboration over contract negotiation. V4. Responding to change over following a plan*	■	■	■	■
Principles	***Outcome Principles.** P1. Our highest priority is to satisfy the customer through early and continuous delivery of valuable software. P7. Working software is the primary measure of progress*	■	■	■	■
	***Project Principles.** P2. Welcome changing requirements, even late in development. Agile processes harness change for the customer's competitive advantage. P3. Deliver working software frequently, from a couple of weeks to a couple of months, with a preference to the shorter timescale*	■	■	■	■
	***Team Principles.** P4. Business people and developers must work together daily throughout the project. P5. Build projects around motivated individuals. Give them the environment and support they need, and trust them to get the job done. P6. The most efficient and effective method of conveying information to and within a development team is face-to-face conversation. P8. Agile processes promote sustainable development. The sponsors, developers, and users should be able to maintain a constant pace indefinitely. P12. At regular intervals, the team reflects on how to become more effective, then tunes and adjusts its behavior accordingly*	■	■	■	■
	***Design Principles.** P9. Continuous attention to technical excellence and good design enhances agility. P10. Simplicity—the art of maximizing the amount of work not done—is essential. P11.The best architectures, requirements, and designs emerge from self-organizing teams*	■	■	■	■

Table 5. Overall assessment of the 8 selected documents on the agile tenets in the agile ITSM methods literature

Type	Tenet	Sem.	Core	Comp.	Factors
Aim	*Providing business value to its customers and users*	■	■	□	□
Values	*V1. The individuals and the interactions are privileged over processes, control activities and tools. V2. The working it services are preferred to comprehensive documentation. V3. The collaboration between employees, as well as with customers and suppliers, takes a prominent place instead of internal and external agreement negotiation. V4. Responding to change is favored compared to following plans*	■	■	□	□
Principles	***Outcome Principles.** P1. The highest priority is to satisfy customers and users through early and constant delivery of valuable and working it services. P7. The primary measure of success has to be the value provided to users by the delivery of quality it services*	■	■	□	□
	***Project Principles.** P2. Welcome changing needs and constraints, whatever the moment, and be structured to harness changes in order to reinforce the customer's competitive advantage. P3. Deliver a useful and warranted IT service working frequently, from a couple of weeks to a couple of months, with a preference to the shorter timescale*	■	■	□	□
	***Team Principles.** P4. Business and technical people must work together daily. P5. Give to the individuals the environment and support they need to be motivated in their daily work and trust them to get the job done. P6. Promote face-to-face conversations to convey information to and within teams. P8. Promote sustainable work by considering that all the it service stakeholders should be able to maintain a constant pace indefinitely. P12. At regular intervals, favour moments of reflection during which teams discuss on how to become more effective, and let them then tune and adjust their behavior accordingly*	■	□	□	□
	***Design Principles.** P9. Promote the simplicity and the technical excellence to create good it service designs, and to implement and support them. P10. Simplicity—the art of maximizing the amount of work not necessary—is essential. P11. Let the teams be self-organized for the design, the implementation, the delivery and the support of it services*	■	■	□	□

4 Discussion and Conclusions

The overall results reported in Table 4 for the agile SwE domain suggest a high degree of adherence between the proffered tenets (aim, values and principles) and the four types of analyzed literature. A common, agreed and shared corpus of agile concepts have been accepted both the agile SwE academic and professional communities. We considered a high conceptual alignment between the aim, values, and principles, and thus they are mutually reinforced between the agile SwE academic and professional communities. Additionally, Table 1 showed clear evidence of the intensity and maturity of the agile SwE research stream, which is lately manifested by a huge community of researchers and practitioners (over 14,000 citations in the 2000–2019 period and an active international agile alliance organization who shares the agile tenets).

In contrast, in Table 5 regarding the ITSM domain, we found that the two emergent ITSM methods, claimed as agile, are fundamentally simplifications from the rigor-oriented and heavy-process ones. FitSM did not include the concept agile in this seminal document [32], and its concept of a lightweight standard should be interpreted differently to agile. For the case of IT4IT [34], in the seminal consulted document, the agile dimension is not directly reported but through the adaption of the DevOps method in the IT4IT method. With this adaption, the IT4IT method approximates conceptually to an agile ITSM paradigm, and to their agile ITSM tenets. However, it can be reported that a common, agreed and shared set of tenets for the agile ITSM domain is still inexistent as it is available in the agile SwE domain. In this study, we used an initial and scarcely known proposal of an agile ITSM framework [16] derived from the agile SwE one [14].

Hence, we concluded that: (1) there is a robust set of tenets in the agile SwE domain; (2) there is a lack of robust tenets in the agile ITSM domain; (3) there is a high adherence on the agile tenets and their corresponding literature in the SwE domain; (4) there is a low one in the agile ITSM domain; and (5) the similarities rely in the transference of the agile tenets from the SwE domain to the ITSM one but with a strong difference in the high level of maturity and adherence between proffered practiced tenets in the agile SwE domain and a low one in the agile ITSM domain.

Hence, thus, a call for a robust theoretical foundation like those found in the agile SwE domain is suggested for the agile ITSM domain, but we considered it will be possible only with the increment in the utilization of these agile ITSM methods by a huge community of practitioners and research be achieved.

Appendix: Statements of Evidence of the Adherence to Agile Tenets from the Agile SwE and ITSM Methods Literature

(See Tables 6, 7, 8 and 9)

Table 6. Analysis of the 12 selected documents on the agile tenets (aim, values) in the agile SwE methods literature

Study	Aim	Values
Seminal	"Rather than planning, analyzing, and designing for the far-flung future, XP exploits the reduction in the cost of changing software" [17] and "to do all of these activities a little at a time, throughout software development" [17]	"XP offered us a unique combination: agility to meet the volatile requirements on time" [17] and "quality to avoid the dreaded rerun" [17]
Core	"Teams using XP are delivering software often and with very low defect rates" [18]. "Cockburn will describe how specific agile practices have been successfully employed in industrial practice" [19]. "Many of the major corporations have announced to pursue for agile solutions with the aim of improving dramatically the lead-times, costs and quality aspects." [20]."individual principles and practices of agile development were… put together into a cogent "theoretical and practical framework" [25]. "Agile has now become a major software engineering discipline in both practice and research." [27]	"Extreme Programming is a discipline of software development based on values of simplicity, communication, feedback, and courage" [18]. "the single largest review (with 333 papers) was on the role of communication, a human and social aspect fundamental to agile software development" [27]
Comp.	"Most view ASD favorably due to improved communication between team members, quick releases and the flexibility of designs in the Agile process." [21]. "four reasons for adopting agile methods: adaptability to change, short time frames of releases, continuous feedback from customers, high-quality and bug free software." [23]	"Documentation and planning are viewed as just-in-time and just-enough for the next sprint." and "Quick Releases" [21]. "we conclude that adoption of agile principles in globalized work is feasible, it has a large potential but may be limited by inherent challenges in distributed work." [22]
Factors	"systems that are developed using agile methods provide only a limited set of functions when they are first introduced to users, but they will evolve periodically" [24]. "Agile methodologies have become well-accepted, with over 65% of companies reporting some type of use of agile methodologies for their software development projects" [28]	"Once users have adapted to the constant changes, they view the changes as a natural part of the system." [24]. "Agile practices are software development tasks or activities that are used to implement the agile method's principles and values." [28]

Table 7. Analysis of the 12 selected documents on the agile tenets (principles) in the agile SwE methods literature

Study	Principles
Seminal	**Outcome Principles.** "Small releases" [17]. "The system is put into production in a few months, before solving the whole problem." [17]. **Project Principles.** Embracing change with Extreme Programming" [17]. "At the scale of weeks and months, you have stories in this iteration and then the stories remaining in this release." [17]. **Team Principles.** "the partners will have a short (15-minute) meeting with the customer" [17] and "and/or with the programmers" [17]. **Design Principles.** "Working with a partner, the programmer makes the test cases run" [17] and "evolving the design in the meantime to maintain the simplest possible design for the system as a whole." [17]
Core	**Outcome Principles.** "The additional process steps, roles, and artifacts helped many teams to enjoy higher success rates" [18] and "more satisfied customers" [18]. **Project Principles.** "these practices have had much success, initially with small teams, working on projects with high degrees of change" [18]. "The principles encourage practices that accommodate change in requirements at any stage" [25]. **Team Principles.** "customers are actively involved in the development process, facilitating feedback" [25] and "reflection that can lead to more satisfying outcomes." [25]. **Design Principles.** "XP teams build software to a simple design" [18]. "They start simple, and through programmer testing and design improvement, they keep it that way." [18]
Comp.	**Outcome Principles.** "reasons for adopting agile methods: adaptability to change, short time frames of releases" [23] and "continuous feedback from customers, high-quality and bug free software." [23]. **Project Principles.** "Quick Releases" [21]. "Flexibility of Design – Quicker Response to Changes" [21]. "Agile methodologies like XP are capable of quickly adapting to the changing requirements of the customers" [23]. **Team Principles.** "Improved Communication and Coordination" [21]. "better communication among the team members" [23]. **Design Principles.** "The top Agile practices that teams followed were team coding standards and continuous integration of code." [21]. "Continuous testing and integration regime …enforce the delivery of high quality bug free software" [23]
Factors	**Outcome Principles.** "comfort with change, faciliting condition and habit are found to be most significant predictors for both intention to continue using agile IS" [24]. "improve software quality" [28]. **Project Principles.** "a supportive infrastructure can promote users' willingness to not only continue using the agile IS but also to use the new features" [24]. "Improve Effectiveness category focuses on adoption motives such as enhancing the organization's ability to manage changing priorities" [28]. **Team Principles.** "increasing productivity, accelerating time to market, and reducing costs" [28]. "Daily Stand Up (Rank 1) refers to a meeting held each day for which every team member attends and provides information to the team" [28]. **Design Principles.** "A standard development cycle was six weeks, and each release contained anywhere from two to three relatively significant enhancements" [24]. "Improved/increased engineering discipline" [28]

Table 8. Analysis of the 8 selected documents on the agile tenets (aim, values) in the agile ITSM methods literature

Study	Aim	Values
Seminal	"ITSM: entirety of activities performed by an IT service provider to plan, deliver, operate and control IT services offered to customers" [32]. "For each service, the service portfolio may include information such as its value proposition" [32]. "IT Value Chain" [34]. "Business value from investments in IT are delivered to the enterprise via an IT operating model" [35]	"The main goal of the FitSM family is to maintain a clear, pragmatic, lightweight and achievable standard" [32]. "that allows for effective IT service management (ITSM)." [32]
Core	"Customer Alignment – the provision of IT services shall be aligned to customer needs." [30]	"the processes around the ITIL framework, specifically, Release Management and to a broader extent, Change Management need to become a little "Agile" in their own way" [29]. "Efficiency and collaboration" and "Strategic Change" [35]
Comp.	N.A.	N.A.
Factors	N.A.	N.A.

Table 9. Analysis of the 8 selected documents on the agile tenets (principles) in the agile ITSM methods literature

Study	Principles
Seminal	**Outcome Principles.** "DevOps is a way of collaborating and industrializing using highly automated approaches to deploy solutions that evolve as fast as your business needs it." [34]. **Project Principles.** "DevOps aims to reduce the impact of changes, to reduce cost, and minimize impact to the live services" [34]. **Team Principles.** "Accommodate the need of self-organizing teams to define their own queues, especially lifecycle states (e.g., "definition of done")" [34]. **Design Principles.** "Services designed for automation and self-service through catalog and APIs" [34]
Core	**Outcome Principles.** "Customer Alignment – the provision of IT services shall be aligned to customer needs." [30]. **Project Principles.** "A big asset of FitSM is the approach to introducing changes in Service Management System. It provides a guidance on what should be done and in what order" [31]. **Team Principles.** N.A. **Design Principles.** "On application side the solution allows much quicker development cycles which will eventually allow products to reach the markets in time to get the right competitive advantage" [31]
Comp.	**Outcome Principles.** "Improvement in user satisfaction" [33]. **Project Principles.** N.A. **Team Principles.** N.A. **Design Principles.** N.A.
Factors	**Outcome Principles.** N.A. **Project Principles.** N.A. **Team Principles.** N.A. **Design Principles.** N.A.

References

1. Hunnebeck, L., et al.: ITIL Service Design. The Stationary Office (TSO), London (2011)
2. Iden, J., Eikebrokk, T.R.: Implementing IT service management: a systematic literature review. Int. J. Inf. Manag. **33**(3), 512–523 (2013)
3. Marrone, M., Gacenga, F., Cater-Steel, A., Kolbe, L.: IT service management: a cross-national study of ITIL adoption. Commun. Assoc. Inf. Syst. **34**, 865–892 (2014)
4. Eikebrokk, T.R., Iden, J.: Strategising IT service management through ITIL implementation: model and empirical test. Total Qual. Manag. Bus. **28**(3–4), 238–265 (2017)
5. Al-Zewairi, M., Biltawi, M., Etaiwi, W., Shaout, A.: Agile software development methodologies: survey of surveys. J. Comput. Commun. **5**(5), 74–97 (2017)
6. Chuang, S.W., Luor, T., Lu, H.P.: Assessment of institutions, scholars, and contributions on agile software development (2001–2012). J. Syst. Softw. **93**, 84–101 (2014)
7. Boehm, B.: Get ready for agile methods, with care. Computer **35**(1), 64–69 (2002)
8. FitSM - a free standard for lightweight ITSM. https://fitsm.itemo.org
9. The Open Group IT4IT™ Reference Architecture, Version 2.1. http://pubs.opengroup.org/it4it/refarch21
10. Smith, A.W., Rahman, N.: Can agile, lean and ITIL coexist? IJKBO **7**(1), 78–88 (2017)
11. Webster, J., Watson, R.T.: Analyzing the past to prepare for the future: writing a literature review. MIS Q., xiii–xxiii (2002)
12. Teddlie, C., Yu, F.: Mixed methods sampling: a typology with examples. J. Mix. Methods. Res. **1**(1), 77–100 (2007)
13. Galvan, S.: Design and Validation of a Project Management Deployment Package for Agile Systems Development Methodologies Based on the ISO/IEC 29110 Standard (Entry Profile). PhD Dissertation, UJAT (2017)
14. Manifesto for Agile Software Development. http://www.agilemanifesto.org
15. Subway Map to Agile Practices. https://www.agilealliance.org/agile101/subway-map-to-agile-practices/
16. Verlaine, B.: Toward an agile IT service management framework. Serv. Sci. **9**(4), 263–274 (2017)
17. Beck, K.: Embracing change with extreme programming. Computer **32**(10), 70–77 (1999)
18. Lindstrom, L., Jeffries, R.: Extreme programming and agile software development methodologies. Inf. Syst. Manag. **21**(3), 41–52 (2004)
19. Ramesh, B., Abrahamsson, P., Cockburn, A., Lyytinen, K., Williams, L.: Agile software development methods: when and why do they work? In: IFIP International Working Conference on Business Agility and Information Technology Diffusion, pp. 371–373. Springer, London (2005)
20. Abrahamsson, P., Still, J.: Agile software development: theoretical and practical outlook. In: 8th International conference on Product-Focused Software Process Improvement, pp. 410–411. Springer, London (2007)
21. Begel, A., Nagappan, N.: Usage and perceptions of agile software development in an industrial context: an exploratory study. In: First International Symposium on Empirical Software Engineering and Measurement, pp. 255—264. IEEE Press, Los Alamos (2007)
22. Hanssen, G.K., Šmite, D., Moe, N.B.: Signs of agile trends in global software engineering research: A tertiary study. In: Sixth International Conference on Global Software Engineering Workshop, pp. 17–23. IEEE Press, Los Alamos (2011)
23. Rao, K.N., Naidu, G.K., Chakka, P.: A study of the Agile software development methods, applicability and implications in industry. Int. J. Softw. Eng. Appl. **5**(2), 35–45 (2011)

24. Hong, W., Thong, J.Y., Chasalow, L.C., Dhillon, G.: User acceptance of agile information systems: a model and empirical test. J. Manag. Inf. Syst. **28**(1), 235–272 (2011)
25. Dingsøyr, T., Nerur, S., Balijepally, V., Moe, N.: A decade of agile methodologies: towards explaining agile software development. J. Syst. Softw. **85**, 1213–1221 (2012)
26. Anand, R.V., Dinakaran, M.: Popular agile methods in software development: review and analysis. Int. J. Appl. Eng. Res. **11**(5), 3433–3437 (2016)
27. Hoda, R., Salleh, N., Grundy, J.: The rise and evolution of agile software development. IEEE Softw. **35**(5), 58–63 (2018)
28. Tripp, J.F., Armstrong, D.J.: Agile methodologies: organizational adoption motives, tailoring, and performance. J. Comput. Inf. Syst. **58**(2), 170–179 (2018)
29. Hoover, C.: ITIL vs. agile programming: is the agile programming discipline compatible with the itil framework? In: International CMG Conference, pp. 613–620. Computer Measurement Group (2006)
30. Radecki, M., Szymocha, T., Szepieniec, T., Różańska, R.: Improving PL-Grid operations based on fitsm standard. In: Bubak, M., Kitowski, J., Wiatr, K. (eds.) eScience on Distributed Computing Infrastructure, pp. 94–105. Springer, London (2014)
31. Komarek, A., Sobeslav, V., Pavlik, J.: Enterprise ICT transformation to agile environment. In: Computational Collective Intelligence, pp. 326–335. Springer, London, 2015
32. FitSM - Part 0: Overview and vocabulary. https://fitsm.itemo.org/wp-content/uploads/sites/3/2018/05/FitSM-0_Overview_and_vocabulary.pdf
33. Melendez, K., Dávila, A., Pessoa, M.: Information technology service management models applied to medium and small organizations: a systematic literature review. Comput. Stand. Interfaces **47**, 120–127 (2016)
34. IT4IT™ Agile Scenario Using the Reference Architecture. https://publications.opengroup.org/w162
35. Campbell, A., Fulton, M.: Defining the IT Operating Model. The Open Group, Berkshire (2017)
36. Michael, S., Michael, B., Thomas, S.: IT service management frameworks compared–simplifying service portfolio management. In: 2019 IFIP/IEEE Symposium on Integrated Network and Service Management, pp. 421–427. IEEE Press, Los Alamos (2019)

Gamification in Software Engineering:
A Tertiary Study

Gabriel Alberto García-Mireles[1]([⊠]) and Miguel Ehécatl Morales-Trujillo[2]

[1] Departamento de Matemáticas, Universidad de Sonora,
Hermosillo, Sonora, Mexico
`mireles@mat.uson.mx`
[2] Computer Science and Software Engineering Department,
University of Canterbury, Christchurch, New Zealand
`miguel.morales@canterbury.ac.nz`

Abstract. Gamification is a research area that influences the extent to which software organizations work and educational practices can be improved in software engineering (SE). Given that several systematic reviews on gamification in SE have been published, this paper aims at understanding how gamification has been addressed, simultaneously identifying its impact on SE. As a result of this systematic mapping, we identified 12 secondary studies. The majority of papers reported the usage of points, badges, and leaderboards as game elements in software engineering process, software engineering methods and tools, and software engineering management. Although secondary studies reported a positive impact of gamification, more empirical research is required. In a nutshell, current research focuses on the feasibility of using gamification in different SE areas; however, there is a need for research of application context and for defining variables under study to carry out more sound empirical research.

Keywords: Gamification · Tertiary study · Software engineering · Systematic mapping

1 Introduction

Gamification is defined as the use of game design elements in non-game contexts [1]. Since 2010, gamification has undergone a very important growth in a number of applications within such domains as finance, health, education, sustainability as well as entertainment media [1, 2]. In the case of software industry, gamification is currently gaining popularity.

In the Software Engineering (SE) field, gamification is a relevant topic of interest since it is applied in both industrial settings [3] and education [4, 5]. In the educational context, Hamari et al. [2] found that majority of empirical studies reported positive effects of gamification, although the studies were more descriptive than based on inferential statistics. Similarly, there are reports that gamification has a positive effect on students' engagement, participation, and increased attendance [6]. In industrial settings, several papers have reported empirical evidence, but there is a need for more

© Springer Nature Switzerland AG 2020
J. Mejia et al. (Eds.): CIMPS 2019, AISC 1071, pp. 116–128, 2020.
https://doi.org/10.1007/978-3-030-33547-2_10

studies [3]. On the other hand, the outcomes of applying gamification are domain specific and highly contextual [7]. Thus, inducing a specific behavior depends on the context. Indeed, understanding the way game mechanics and dynamics impacts on motivation and engagement requires years of empirical research [7].

Currently, technical, cultural, economic and political forces demand a design practice based on computing technology for enhancing user experience and engagement in different domains and industries [8]. There is a research shift from asking whether gamification works to a more mature research in terms of how to design gamification, to explain the underlying mechanisms that make gamification work, and the differences in individuals and context that point towards the dependence of a gamification design of the application context [8].

Thus, the aim of this paper is to identify current trends in SE research about gamification as a first step to assess its research focus. Given that several systematic reviews and mapping studies have been published (e.g. [3–5]), we decided to conduct a tertiary study [9] to identify secondary studies that address gamification in SE. From the 12 selected studies, we found that there is a general positive gamification effect in the studied contexts, but little evidence is reported on the rigor with which research has been conducted. Thus, the analysis of the secondary studies suggests that gamification research in SE is currently trying to answer the question whether gamification actually works in SE.

The remainder of this paper is structured as follows. Section 2 describes relevant tertiary studies that support the endeavor of this review. Section 3 details the methodology followed to conduct the systematic mapping. Section 4 describes main outcomes of the study, while Sect. 5 presents discussion. Finally, Sect. 6 presents conclusions and suggestions for further research.

2 Related Work

A tertiary study is "a review of secondary studies related to the same research question" [9]. A tertiary study is a useful map to introduce people to a research area (e.g., PhD students and practitioners) who are interested in it [10]. In the field of SE, some tertiary studies have been published. For instance, Garousi and Mantyla [10] reported finding no tertiary studies about software testing, which was the main topic of their research, however they found a handful number of tertiary studies, which were reviewed by the authors in order to identify similarities of research settings.

Similarly, Rios et al. [11] reported that 25 tertiary studies have been conducted in SE. An important research area in this type of research is the quality of secondary studies in SE. Despite several topics are addressed in these tertiary studies, studies about technical debt were not found. Based on [10, 11], we conducted the research presented in this study.

Following the same search strategy of aforementioned papers, we looked for tertiary studies on gamification in SE, however, none were found. The search string used was: (gamification OR gamified OR gamifies OR gamify OR gamifying) AND "tertiary study"). It was executed in Scopus, Web of Science, IEEE Xplore, and the ACM digital library. Not a single tertiary study was retrieved from these digital databases on June 17, 2019.

3 Conducting the Tertiary Study

The general goal of this study is to identify systematic reviews about gamification in SE. Since targeted papers in this review are secondary studies, our review is a tertiary study. A secondary study is "A study that reviews all the primary studies relating to a specific research question with the aim of integrating/synthesizing evidence related to a specific research question" [9]. The review method is based on guidelines for conducting systematic reviews [9] and mapping studies [12]. This section presents main aspects of the research protocol.

3.1 Research Questions

A general research question stated in this study is: What is the evidence of gamification research in SE as reported in secondary studies? We decided to conduct a tertiary study about gamification in SE due to the fact that a tertiary study is able to gather current best evidence on this topic. We found that several secondary studies call for more and better quality in research for primary papers [3, 5]. Thus, it is relevant to understand the topics addressed in current secondary studies as well as the extent quality of primary studies are addressed. In order to answer the general research question, we formulated the following specific research questions:

- RQ1. What systematic reviews have been published in the area of gamification in SE?
- RQ2. What research questions have been addressed?
- RQ3. What evidence is reported to support research directions and practical knowledge?
- RQ4. What approaches are used to identify and assess empirical studies?

The purpose of RQ1 is to identify secondary studies that have addressed gamification in SE and their general publication trends (year, venue, SE knowledge areas, and quality assessment). Question RQ2 explores the main research questions and key constructs studied whereas RQ3 focuses on the evidence (positive and negative) reported in secondary studies as well as challenges identified. Finally, RQ4 identifies empirical studies within each one secondary study and how their quality was assessed.

Research questions helped to define the search string, which is focused on gamification in SE. Thus, as based on PICO approach [9], the population is the papers which address gamification in SE while the intervention correspond to different types of systematic reviews that have been conducted. Both comparison and outcome are not considered in the formulation of the search string because this tertiary study aims at describing general results, but without comparing findings of secondary papers. The search string has three main concepts: gamification, software engineering, and systematic review (Table 1) [13]. The list of synonyms for gamification was based on [5], while relevant terms for SE knowledge areas were taken from SWEBOK [14]. The list of alternative terms for systematic review was adapted from [15]. The search string, Gamification AND Software Engineering AND Systematic Review, was adapted to be executed in Scopus, Web of Science, IEEE Xplore, and ACM Digital Library.

Table 1. Terms used in the search string.

Main term	Alternative terms
Gamification	gamification OR gamified OR gamifies OR gamify OR gamifying
Software Engineering	"software engineering" OR "software requirements" OR "software design" OR "software construction" OR coding OR "software testing" OR "software quality" OR "software engineering process" OR "software process" OR "software engineering management" OR "software project" OR "software maintenance"
Systematic Review	"systematic literature review" OR "systematic review" OR "research review" OR "research synthesis" OR "research integration" OR "systematic overview" OR "systematic research synthesis" OR "integrative research review" OR "integrative review" OR "mapping study" OR "scoping study" OR "systematic mapping"

3.2 Selection of Studies and Data Extraction

The selection criteria for including secondary studies are as follows:

1. Peer-reviewed papers about systematic reviews or mapping studies reporting that a method for conducting systematic reviews was followed.
2. The topic of the systematic review or mapping study is gamification in any of the SE knowledge areas.
3. The report must include research questions, research process, data extraction, and a classification, summary or synthesis of results.

   Exclusion criteria are the following:

1. Ad hoc literature reviews that lack defined research questions, selection process or extraction process.
2. Duplicate papers where authors report the same study, i.e., same objective or research questions, or one of them is an extended version of the other. Thus, most complete or updated version will be selected.

The procedure for selecting secondary studies was carried out as follows. The first author executed the search string in Scopus, which retrieved 26 records on March 28, 2019. Both title and abstract were read from each retrieved record to select the candidate secondary studies by checking that they address the topic under study. The candidate list of 15 papers included several duplicates (both first and extended version of a study). This list was used to eliminate duplicates in the outputs gathered from executing the search in Web of Science (14 records), IEEE Xplore (8 records), and ACM Digital Library (3 records). A full reading of papers in the candidate list was carried out to verify they meet selection criteria. As a result, 12 secondary papers were selected. A snowballing procedure was executed to identify other studies. Backward snowballing was applied to review 443 references from the 12 secondary papers and the forward snowballing considered the 162 cites available in Scopus on June 17, 2019. The snowballing procedure did not add any paper. The automatic procedure for selecting papers was verified by the second author.

A template for extracting data was designed considering the research questions. Following data was extracted from included studies: reference, publication year, type of publication (journal/conference), type of secondary study as reported by authors, study objective, research questions, search term used, databases searched, year covered, gamification topic discussed, SE area, guidelines for conducting systematic reviews, summary of findings and quality score. Extracted data was reviewed by the second author, differences found were resolved by consensus.

3.3 Quality Assessment

Quality assessment was conducted to determine transparency of the process and the trustworthiness of conclusions provided in each systematic review. The quality of each study was assessed according to the DARE method [9], which considers the following criteria:

QC1. Inclusion and exclusion criteria are described and appropriate.
QC2. Literature search is likely to have covered all relevant studies.
QC3. Reviewers assess the quality of the included studies.
QC4. Basic data of studies is adequately described.

Each criterion is scored with the values 1, 0.5 and 0. For QC1, 1 is assigned when both inclusion and exclusion criteria are explicitly described in the review. The value 0.5 is assigned when only one selection criteria is described or they are implicit. The value 0 is used when selection criteria is not defined. For QC2, the value 1 is assigned when at least 4 digital libraries were searched and additional search strategy was used. The value 0.5 is assigned when search in only 3 to 4 digital libraries is reported. The value 0 is assigned when the paper reports up to 2 digital libraries or a small set of journals.

For QC3, the value 1 corresponds to the paper which explicitly defines the quality assessment of the empirical evidence provided in each primary study. The value 0.5 is assigned when the assessment is focused on answering the research question posed in the primary paper. The value 0 is assigned when there is no explicit quality assessment of primary papers. For QC4, the value 1 corresponds to articles that present data about each primary study. The value 0.5 is assigned when primary studies' data is presented in a summarized fashion. The value 0 is assigned when data of the individual primary studies are not specified.

4 Systematic Reviews About Gamification in SE

In this section, we will present the main findings of this tertiary study. The results are depicted following the same order of the research questions.

4.1 Systematic Reviews in Gamification in SE

As a result of conducting the tertiary study, we found that 12 secondary studies met the selection criteria. Two of them were published in 2015 while 6 in 2018 (Fig. 1). The list of papers is presented in Appendix A. Considering the venue of the article; four of them were published in journals (S05, S06, S09, and S12) and the remaining eight, in conferences.

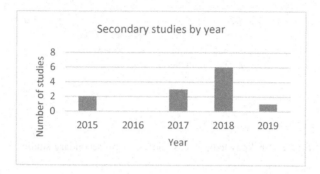

Fig. 1. Publication trends of systematic reviews about gamification in SE.

The secondary studies address software development process (Fig. 2) followed in both SE courses and in industry (S05, S08, S10, and S12). S05 investigates evidence of using gamification in SE educational settings; and it studies software processes that have been gamified. S08 presents a set of primary papers which address gamification of software process improvement approaches. Similarly, S10 investigates application of game elements in software development in both research and industrial levels while S12 reviews fields of gamification in SE and characterizes them by using the ISO/IEC 12207 processes.

Software engineering models and methods knowledge area are addressed in two papers (S03, S07). S03 identifies different gamification design framework reported in literature while S07 reviews studies that support monitoring of gamification processes. Concerning the software engineering management area, S01 studies gamified processes while S09 investigates gamification in the context of both creating and improving the performance of work teams. In the engineering foundations area, S06 investigates game-based methods used in SE education. From 156 identified papers, 10 were categorized in a gamification approach (S06). Also there are reviews about software quality (S11), software requirements (S02), and software testing (S04).

With regard to the quality of each secondary study, we applied the criteria mentioned in Sect. 3.3 Quality Assessment. Figure 3 depicts a frequency of each quality score. The lowest value, 1.5 was gathered by one paper (S08) while the maximum value, 4, belongs to two papers (S02, S12). The remaining papers gathered 2 (S03, S04, S10), 2.5 (S01, S06, S09), and 3 (S05, S07, S11).

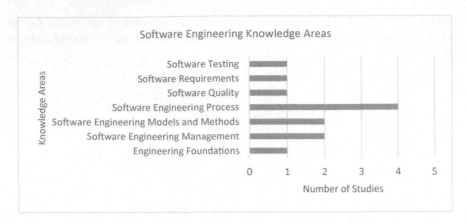

Fig. 2. SE knowledge areas addressed by secondary studies.

4.2   Research Questions Addressed

All reviewed papers present from two (S07, S08) to nine research questions (S05) where the mode is five questions (S02, S06, S09, S12). In total, 59 questions are mentioned. Besides, questions are presented in two formats. Some research questions are presented after establishing the aims of the study (S01, S06, S07, S08, S09, S10) while other studies posed a hierarchy of general-specific question format. In the latter category, five studies posed only one general question (S02, S03, S04, S11, and S12) with several specific research questions, and one paper presented three general questions, two of them with additional specific questions (S05).

Fig. 3.   Quality assessment of selected studies.

In order to classify the topics addressed by research questions, each question was extracted and labeled concerning a particular aspect addressed. The labelling allowed classifying questions into the following categories: application area, gamification elements, gamification evidence, and research approach. Application area includes questions related to gamified domain (7), and contextual factors considered in the domain (11) such as industrial context (1). Gamification elements category is decomposed into gamification mechanics and dynamics (11), methods for designing and implementing gamification (8) and gamification examples (1). Concerning gamification evidence, this category includes gamification goal questions (4), impact of gamification elements (1), gamification issues (1), and gamification conclusions such as benefits, drawback, and challenges, among others (3). The category of research approach consists of codes for research methods applied (4), state of the art (6) and type of publications (1).

4.3 Evidence About Gamification in SE

Different gamification elements have been reported in secondary studies (Table 2). S02 reports 25 different gamification elements while S11 found 15. Points, leaderboards, badges and levels are the most frequently reported gaming components. Rewards, feedback, challenges, and competition are some game mechanics reported. Progression also is reported as game dynamics. Thus, there is a focus on reporting game components (points, levels, leaderboards, and badges) with little attention to mechanics and dynamics. In addition, S05 noticed that few papers clarify the approach used to select gamification elements.

The application of gamification practices is yet immature. Majority of proposals are limited to theoretical aspects and few experiments. Besides few solutions are implemented in industrial settings and majority of studies are limited to a first prototype to show the effect of the concept (S03). Besides, S12 observed that in more than half of their studies, points and badges are the game components implemented, however, others reported that leaderboards and badges are the most important game component. Focusing only on game components, without an appropriate consideration of other game elements (e.g., mechanics and dynamics), could produce the failure of gamified applications due to a poor design [3].

Table 2. Most common game elements reported.

Game elements	Studies
Points	S01, S02, S04, S05, S06, S09, S10, S11, S12
Level	S01, S04, S05, S09, S10, S11
Leaderboards	S01, S02, S04, S05, S06, S09, S10, S11
Badges	S01, S02, S06, S09, S10, S11, S12
Rewards	S01, S05, S10, S11
Progression	S05, S09

Other aspect that should be considered in designing gamification methods is the application context. S12 found that a common strategy in gamification is to incorporate a new tool that is developed ad-hoc. However, they believe that the tool should be integrated into the organization's infrastructure in order to be successful. In the educational context, proposals are based on the use of a new web platform, gamified application, or gamification based on learning management systems (S09, S05).

Gamification has been studied in industry (e.g., S01, S09, S10, S12) and education (e.g., S05, S08). S11 reported that gamification, in the context of software quality, is most investigated in education and most of the papers deal with gamified courses [16]. Indeed, S06 observed two strategies in gamification in SE education: gamification of the classroom experience and gamification of specific SE activities. Within software processes, software requirements, software development, software testing, and software project management are the areas that attracted the greatest interest in the field of gamification in SE (S12).

Gamification is used because it can influence the behavior of users by means of users' engagement and motivation [2]. Among the reported change in psychological or behavior of users of gamified systems and applications, the most mentioned are engagement (S02, S03, S04, S05, S06, S09, S10, S11) and motivation (S02, S03, S04, S06, S08, S09, S10, S11). Other variables, related to participants' behavior, are improving communication, collaboration and social skills (S02, S01, S05, S09, S10) and improving technical skills (S04, S05, S06, S09). Besides, others behaviors observed are related to eliciting more innovative requirements (S02), enhancing the quality of software (S04), using best practices (S05, S06), doing assigned activities (S09), applying gamification as a goal (S09), improving code reviews (S09), participating in software process improvement activities (S09), enhancing participants' performance (S10) of software development actors and increasing the quality of development process (S10). In addition, it is reported satisfaction with gamified software (S11), and enhancements on effectiveness (S11), usability (S11) and functional suitability (S11). However, few papers discuss the specific definitions of terms used (e.g. engagement has different meanings as well as motivation).

Conclusions of secondary studies noted a general positive impact of gamification in SE (S01) in terms of improved productivity and a generation of shared context among participants, and also enhancing quality of software product and increasing the performance of developers (S10). However, others studies reported both few primary papers (S02) and low quality studies to gather conclusive results about the effect of gamification (S03, S05, S12). Indeed, existing research on gamification in SE is very preliminary (S12). Other aspects reported are related to the little use of the variety of gamification elements (S02, S12) and the need to address quality aspects in gamified applications (S03) since given solutions still present shortcoming in sustainability, balancing the work and ethics or the lack of consideration of several quality characteristics of a software product (S11). Other aspect that requires attention is related to the difficulties that are found in the specific context in which gamification is used (S05, S06, S07, S12).

4.4 Identification of Empirical Studies

From the 12 secondary studies, five are categorized considering the title and abstract as 'systematic literature reviews' (S01, S02, S07, S09, S10) while the remaining studies are systematic mappings or systematic mapping studies (seven studies). While systematic mapping studies do not require assessment of the quality of primary papers [17], systematic literature reviews should discuss the quality assessment of primary papers [18]. Thus, four systematic literature reviews (S01, S02, S07, S09) consider criteria to assess quality of primary papers. Each of studies S01, S07 and S09 include three questions to assess primary studies, which are related to the topic under study. Study S02 evaluates quality of primary studies through ten questions, where two of them are related to the description of both goals of empirical research and validity threats. Considering the mapping studies, S12 uses assessment criteria of five questions. As a whole, this paper (S12) reports that empirical evidence is lacking. Thus, quality assessment is poorly reported in the secondary studies.

Wieringa et al. [19] classification supports identification of empirical papers. Within the set of selected papers, six classify the type of research (S01, S03, S05, S08, S09, S12) and include both validation and evaluation research papers, which requires the paper to describe an empirical inquiry [17]. The number of primary papers reported for gamification in SE is in the range from 10 to 58 (S06 reports 10 for gamification approach). Empirical studies (validation + evaluation categories from [19]) are reported in the range from two to 32. In average, 50.7% of reported studies are empirical and their range is between 15 and 91%. Thus, it is relevant to understand to what extent the empirical research supports claims of each study.

5 Discussion

This tertiary study presents the research trends about the way systematic reviews and mapping studies have addressed gamification in SE. With regard to the first research question, RQ1, we identified 12 secondary studies (Appendix A) which were published from 2015 to 2019. Half of them were published in 2018. SE knowledge areas addressed, with more than one secondary study, were software engineering process, software engineering models and methods, and software engineering management. A third of secondary studies were published in journals and the quality assessment of 9 of the secondary studies belongs to the range 2.0 to 3.0. The main issue about quality assessment is the lack of information about the quality of the empirical studies included in each secondary study.

Concerning RQ2, we found that 59 questions were posed in selected secondary studies. The topics addressed in them were categorized as application area, gamification elements, gamification evidence, and research approach. A relevant number of questions focused on the context and the domain were gamification was implemented. Gamification elements also are an important category of questions, several focused on gamification components and methods for designing/implementing a gamification solution. Remaining categories has less than ten questions each one (Sect. 4.2 Research Questions Addressed).

About the evidence of gamification in SE, RQ3, we found that points, leaderboards, levels and badges are the most common gamification elements reported. The majority of papers mentioned engagement and motivation as main variables to consider in a gamified application, but their respective definition is lacking. However, the application of gamification practices is still immature. It is required to consider the application context in the design of a gamification solution and to conduct more empirical research.

With regard to the identification and quality assessment of empirical papers, we found that systematic literature reviews poorly address the quality of their respective primary studies. From 215 primary papers reported in seven studies, 109 of them were identified as empirical. However, the majority of them were not assessed with regard to their research rigor.

To mitigate validity threats, we based this tertiary study in sound procedures [9, 12]. In order to mitigate the impact of selection bias, a protocol was built in the first stages of this research. In this document, the key search terms were identified and alternative terms were included. The selection of papers was carried out by one researcher and the other verified the list of selected papers. With regard to the publication bias, it was mitigated by considering four databases common in software engineering field [20] and carrying out a snowballing procedure. Finally, to mitigate the bias due to extraction procedures, a template was built to extract verbatim data from each selected paper.

6 Conclusions

In this tertiary study a set of secondary papers addressing gamification in SE was identified. The general impact of gamification in both industry and education is positive, but few gamification elements have been implemented. The most common are points, levels, leaderboards and badges. The knowledge areas studied are software engineering process, software engineering management and software engineering methods and tools. However, little empirical research is reported and few secondary studies assessed the quality of primary studies. Indeed, it is required that more sound empirical research is conducted to determine the extent gamification elements influence motivation, engagement and behavior of participants. Besides, systematic approaches for incorporating gamification are required to appropriately address the contextual settings.

Given that a main issue is the rigor with which research is reported, there is a need to identify current practices for reporting empirical research in order to develop a set of guidelines. Other topic that is under our consideration is the development of systematic approaches to introduce gamification elements.

Appendix A: Selected Secondary Studies

The list of selected secondary studies included in this Section depicts the study id, the reference, publishing year, venue, software engineering knowledge area (SE KA) and the number of primary studies reported.

sID	Year	Venue	SE KA	Primary papers
S01 [21]	2019	Conference	SE Management	49
S02 [22]	2018	Conference	Software Requirements	8
S03 [23]	2018	Conference	SE Models and Methods	58

(continued)

(*continued*)

sID	Year	Venue	SE KA	Primary papers
S04 [24]	2018	Conference	Software Testing	15
S05 [5]	2018	Journal	SE Process	21
S06 [4]	2018	Journal	Engineering Foundations	156 (10 papers address gamification)
S07 [25]	2018	Conference	SE Models and Methods	2
S08 [26]	2017	Conference	SE Process	13
S09 [27]	2017	Journal	SE Management	31
S10 [28]	2017	Conference	SE Process	12
S11 [16]	2015	Conference	Software Quality	35
S12 [3]	2015	Journal	SE Process	29

References

1. Deterding, S., Dixon, D., Khaled, R., Nacke, L.: From game design elements to gamefulness: defining gamification. In: Proceedings of the 15th International Academic MindTrek Conference: Envisioning Future Media Environments, pp. 9–15. ACM (2011)
2. Hamari, J., Koivisto, J., Sarsa, H.: Does gamification work?–A literature review of empirical studies on gamification. In: 2014 47th Hawaii International Conference on System Sciences (HICSS), pp. 3025–3034. IEEE (2014)
3. Pedreira, O., García, F., Brisaboa, N., Piattini, M.: Gamification in software engineering–A systematic mapping. Inf. Softw. Technol. **57**, 157–168 (2015)
4. Souza, M.R.A., Veado, L., Moreira, R.T., Figueiredo, E., Costa, H.: A systematic mapping study on game-related methods for software engineering education. Inf. Softw. Technol. **95**, 201–218 (2018)
5. Alhammad, M.M., Moreno, A.M.: Gamification in software engineering education: a systematic mapping. J. Syst. Softw. **141**, 131–150 (2018)
6. Dicheva, D., Dichev, C., Agre, G., Angelova, G.: Gamification in education: a systematic mapping study. J. Educ. Technol. Soc. **18**(3), 75–88 (2015)
7. Helmefalk, M.: An interdisciplinary perspective on gamification: mechanics, psychological mediators and outcomes. Int. J. Serious Games **6**(1), 3–26 (2019)
8. Nacke, L.E., Deterding, C.S.: The maturing of gamification research. Comput. Hum. Behav. **71**, 450–454 (2017)
9. Kitchenham, B., Charters, S.: Guidelines for performing systematic literature reviews in software engineering. In: Technical report EBSE-2007-01 (2007)
10. Garousi, V., Mantyla, M.V.: A systematic literature review of literature reviews in software testing. Inf. Softw. Technol. **80**, 195–216 (2016)
11. Rios, N., Neto, M.G.M., Spínola, R.O.: A tertiary study on technical debt: types, management strategies, research trends, and base information for practitioners. Inf. Softw. Technol. **102**, 117–145 (2018)
12. Petersen, K., Feldt, R., Mujtaba, S., Mattsson, M.: Systematic mapping studies in software engineering. In: EASE 2008, pp. 1–10. BCS eWIC (2008)
13. Budgen, D., Brereton, P., Drummond, S., Williams, N.: Reporting systematic reviews: some lessons form a tertiary study. Inf. Softw. Technol. **95**, 62–74 (2018)

14. Bourque, P., Fairley, R.E.: Guide to the software engineering body of knowledge (SWEBOK (r)): version 3.0. IEEE Computer Society (2014)
15. García-Mireles, G.A., Moraga, M.A., García, F., Piattini, M.: Approaches to promote product quality within software process improvement initiatives: a mapping study. J. Syst. Softw. **103**, 150–166 (2015)
16. Vargas-Enriquez, J., Garcia-Mundo, L., Genero, M., Piattini, M.: A systematic mapping study on gamified software quality. In: 7th International Conference on Games and Virtual Worlds for Serious Applications (VS-Games), pp. 1–8 IEEE (2015)
17. Petersen, K., Vakkalanka, S., Kuzniarz, L.: Guidelines for conducting systematic mapping studies in software engineering: an update. Inf. Softw. Technol. **64**, 1–18 (2015)
18. Cruzes, D., Dyba, T.: Research synthesis in software engineering: a tertiary study. Inf. Softw. Technol. **53**, 440–455 (2011)
19. Wieringa, R., Maiden, N., Mead, N., Rolland, C.: Requirements engineering paper classification and evaluation criteria: a proposal and discussion. Requir. Eng. **11**, 102–107 (2006)
20. Kitchenham, B., Brereton, P.: A systematic review of systematic review process research in software engineering. Inf. Softw. Technol. **55**(12), 2049–2075 (2013)
21. Machuca-Villegas, L., Gasca-Hurtado, G.P.: Gamification for improving software project management processes: a systematic literature review. In: Mejia, J., Muñoz, M., Rocha, Á., Peña, A., Pérez-Cisneros, M. (eds.) Trends and Applications in Software Engineering. CIMPS 2018. Advances in Intelligent Systems and Computing, vol. 865, pp. 41–54. Springer, Cham (2019)
22. Cursino, R., Ferreira, D., Lencastre, M., Fagundes, R., Pimentel, J.: Gamification in requirements engineering: a systematic review. In: 11th International Conference on the Quality of Information and Communications Technology (QUATIC), pp. 119–125. IEEE (2018)
23. Azouz, O., Lefdaoui, Y.: Gamification design frameworks: a systematic mapping study. In: 2018 6th International Conference on Multimedia Computing and Systems (ICMCS), pp. 1–9. IEEE (2018)
24. de Jesus, G.M., Ferrari, F.C., de Paula Porto, D., Fabbri, S.C.P.F.: Gamification in software testing: a characterization study. In: Proceedings of the III Brazilian Symposium on Systematic and Automated Software Testing, pp. 39–48. ACM (2018)
25. Trinidad, M., Calderón, A., Ruiz, M.: A systematic literature review on the gamification monitoring phase: how SPI standards can contribute to gamification maturity. In: Stamelos, I., O'Connor, R., Rout, T., Dorling, A. (eds.). Software Process Improvement and Capability Determination. SPICE 2018. Communications in Computer and Information Science, vol. 918, pp. 31–44. Springer, Cham (2018)
26. Gómez-Álvarez, M.C., Gasca-Hurtado, G.P., Hincapié, J.A.: Gamification as strategy for software process improvement: a systematic mapping. In: 2017 12th Iberian Conference on Information Systems and Technologies (CISTI), pp. 1–7. IEEE (2017)
27. Hernández, L., Muñoz, M., Mejía, J., Peña, A., Rangel, N., Torres, C.: Una Revisión Sistemática de la Literatura Enfocada en el uso de Gamificación en Equipos de Trabajo en la Ingeniería de Software. RISTI - Revista Ibérica de Sistemas e Tecnologias de Informação **21**, 33–50 (2017). https://dx.doi.org/10.17013/risti.21.33-50
28. Olgun, S., Yilmaz, M., Clarke, P.M., O'Connor, R.V.: A systematic investigation into the use of game elements in the context of software business landscapes: a systematic literature review. In: Mas, A., Mesquida, A., O'Connor, R., Rout, T., Dorling, A. (eds.) Software Process Improvement and Capability Determination. SPICE 2017. Communications in Computer and Information Science, vol 770, pp. 384–398. Springer, Cham (2017)

Knowledge Management

Knowledge Transfer in Software Companies Based on Machine Learning

Jose Luis Jurado[1]([⊠]), Alexandra CastañoTrochez[1], Hugo Ordoñez[1], and Armando Ordoñez[2]

[1] Universidad de San Buenaventura, Cali, Colombia
{jljurado,hordonez}@usbcali.edu.co,
ctalexandra95@gmail.co
[2] University Foundation of Popayán, Popayán, Colombia
jaordonez@unicauca.edu.co

Abstract. Innovation is a key driver of competitiveness and productivity in today's market. In this scenario, Knowledge is seen as a company's key asset and has become the primary competitive tool for many businesses. However, An efficient knowledge management must face diverse challenges such as the knowledge leakage and the poor coordination of work teams. To address these issues, experts in knowledge management must support organizations to come up with solutions and answers. However, in many cases, the precision and ambiguity of their concepts are not the most appropriate. This article describes a method for the diagnosis and initial assessment of knowledge management. The proposed method uses machine-learning techniques to analyze different aspects and conditions associated with knowledge transfer. Initially, we present a literature review of the common problems in knowledge management. Later, the proposed method and its respective application are exposed. The validation of this method was carried out using data from a group of software companies, and the analysis of the results was performed using Support Vector Machine (SVM).

Keywords: Knowledge management · Knowledge transfer · Artificial Intelligence · Machine learning

1 Introduction

Knowledge has become one of the key assets of companies since the 90 s. Thus, proper knowledge management may support process improvement and create competitive advantages in dynamic markets. This situation is especially true in the software industry, where technology, paradigms, programming languages, project management techniques change rapidly [1].

Knowledge management must face significant challenges in today's software industry. First, the globalization and the widespread use of the Internet, have paved the way for online work. Thus, collaborators from all around the world become sources of knowledge [2]. Manage this sort of knowledge is challenging due to cultural issues and knowledge losses. This problem can be seen, for example, in open source communities where there is a constant and uncontrolled turnover of personnel [3].

© Springer Nature Switzerland AG 2020
J. Mejia et al. (Eds.): CIMPS 2019, AISC 1071, pp. 131–140, 2020.
https://doi.org/10.1007/978-3-030-33547-2_11

Another common problem to achieve an effective knowledge transfer is the loss of information. Most of the employees' knowledge is obtained from sources such as documents, evidence, internet information, repositories among others, called as explicit knowledge, however, the tacit knowledge that resides in each person because of their experience is not adequately recovered.

Several approaches can be found for improving knowledge management. Most of them focus on social behaviors and problems that affect the transfer of tacit knowledge. Some examples are the interaction techniques such as the TMS (transactional memory) [4], gamification techniques or graphs, and correlation matrix [5]. Moreover, the use of the internet of things has been proposed to improve the flow and management of knowledge [6]. Questionnaires are made(with 210 questions, 45 based on the work scale) to relate self-control and mandatory relations between employees with the impact on the tacit exchange of knowledge [7].

By its part, the information collected by companies is a valuable source of analysis to find the deficiencies in knowledge transfer. Currently, experts on knowledge management carry out this analysis by identifying, evaluating, and exposing a solution for the mitigation of problems in the knowledge transfer. However, Machine learning is a promising tool for solving this sort of problems. An example is the Know-Net-Soft project in which they use the K-means algorithm to cluster and identify who are the experts in the company [8]. The contribution generated by this research in the field of knowledge management is the use of SVM classification techniques to determine an applied research in the analysis of problems in knowledge management.

This article explores the possibility of using an automatic learning technique within a reference method, for the diagnosis of the current situation of knowledge transfer. The approach is validated with a set of software companies. To this purpose, some instruments are used to define variables considered part, or source of the leakage. Equally, some instruments are used to analyze the coordination of teams.

2   Theoretical Background

This section describes some concepts associated with knowledge transfer and machine learning.

2.1   Knowledge Transfer

Knowledge management refers to the use of collective knowledge of organizations to achieve sustainability. The dynamic model of knowledge management is composed of 4 concepts. First, socialization (knowledge is tacit and is acquired through observation, imitation, and practice). Second, externalization (Tacit knowledge becomes explicit through dialogue and collective reflection, this process uses metaphors, concepts, hypotheses, among others). Third, internalization (Conversion of the explicit to the tacit through experience, perspective and others). Fourth, and, combination (transforms explicit knowledge into improved explicit, also it is about the transformation of the conceptualization adapted to the current environment, improves accessibility to knowledge) [9].

In knowledge management models, knowledge transfer is the bridge between the creation of knowledge and transformation to explicit knowledge [10]. For [11], Knowledge transfer allows reusing existing experience and can be achieved by recording and processing personal and organizational experience from software development projects [12]. Similarly, for "The transfer of knowledge in organizations is the process by which a unit (for example, a group, a department or a division) is affected by the experience of another" [13]. They proposed the theory of knowledge creation and identified two types of knowledge. The first is the explicit knowledge defined as that which is transmitted in formal and systematic languages. This knowledge can be articulated to include grammatical statements, mathematical expressions, among others. Due to these characteristics, explicit knowledge is easily transmissible between individuals. The second type is tacit knowledge and refers to personal knowledge in a specific context.

2.2 Machine Learning

Machine learning is defined as "The science and art of programming computers so they can learn from data" [14]. Machine learning seeks to identify patterns in the data that support decision making under uncertainty [15].

Machine learning can be supervised or unsupervised. In the first, the agent starts from a set of training examples previously labeled according to a category (for example a set of emails labeled as spam or not spam). The agent classifies new examples based on the training set (in the example, the agent compares the new emails with the ones already labeled). In unsupervised machine learning, the agent classifies the data into different groups without prior information [22].

Support Vector Machine (SVM): First developed by [16] for classification problems, SVM attempts to find the optimal separating hyperplane among the classes (categories) by maximizing the margin between them. Here, points lying on the boundaries are referred to as support vectors, while the middle of the margin is called the optimal separating hyperplane. This margin maximization property of SVM leads to its state-of-the-art classification performance [17]. SVM is particularly suitable for use with large data sets, that is, those with a large number of predictive fields [18].

3   Problems in Knowledge Transfer

The main issues found in the literature are knowledge leaks and team coordination.

Knowledge Leak: is generated during the transformation of tacit to explicit knowledge within knowledge networks. Authors such as [4] have identified factors such as cultural discrepancy, ambiguity in language translations, diversity of data, considerable time, and expense in identifying relevant knowledge within organizations. Most of the problems are related to the knowledge of human resources, communication, and risks [19]. Thus, cultural understanding in the teams is fundamental for both the client and the provider. Moreover, in the open source communities, the high turnover is a significant challenge [3].

Team Coordination: is the organizational capital of the company and is part of the intangible assets along with human capital and information. When these three aspects are united, the organization mobilize and sustain the processes of change to execute its strategy [20]. The creation of a shared mind is generated in formal or informal groups or teams through which the know-how and sense of doing are shared and drives the change of behavior patterns [21].

For the coordination of teams, in [4], problems were identified in the culturally remote work units, as well as the difficulty in the transfer of knowledge from one unit to another, the participants have insufficient background information among themselves and lack a shared language and common interests, which significantly limits their ability to evaluate and share knowledge. Coordination of teams can be affected by three factors: Specialization of personnel, credibility between work groups, and coordination of individual knowledge. The authors [7], highlight in the results of their study, the relationship between employees, having key factors such as a mandatory bond, emotional bond, and self-control. In addition to this, in the case of [22], the problem of low motivation on the part of the teams for the generation of knowledge was described.

4   Description of Method Proposed

The proposed method aims at identifying the problems in knowledge transfer, namely, knowledge leakage and team coordination. Figure 1 shows the three steps of the method that are explained in the following sections.

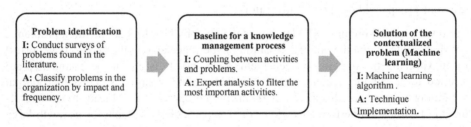

Fig. 1. Phases of the proposed method, each one comprising the name, the instrument (I) and the activity (A).

Problem Identification: In the first step, a systemic review of the literature is carried out on the issues that affect Knowledge transfer, such as leakage of knowledge and teams coordination. Subsequently, with the above information and through instruments such as surveys or elicitation techniques, a set of questions are applied to obtain contextualized information about the company.

The deliverable of this stage is the report with the results of the surveys. This report shows the aspects associated with the problems of the process, and the internal variables of these aspects, which are identified for each business model.

A Baseline for a Knowledge Management Process: here, the activities of the process with a more significant impact on collaborative processes are identified, as well as the roles, tasks, and responsibilities in the processes. Examples of these tasks can be knowledge networks and knowledge selection to transfer.

An expert must carry out this step; he must match the activities with the information obtained in step 1 and classify the data of the problem according to the activities. The purpose of this step is to analyze the issues in the process profoundly and review the results of step one to find the activities that can be applied to solve these problems.

Table 1. Description of the variables selected as sources of knowledge leakage and problems in the coordination of teams.

Variable dictionary	
Decoupling	The employees leave the company without sharing the knowledge obtained adequately
Language comprehension	People who are of different native language so the interpretation can change
Culture_work	Different ways of working, including work standards and the person's own culture
Disinterest	Employees are not interested in doing additional work to the one they have and are not interested in socializing what they have learned
Incomprensión dialect	People from different cities that understand the language but uses different dialects
Uncomprehensible sources	The documentation is not formal at all, does not have a standard; the terminology used is not understandable, nor is in the way in which it was expressed by the person who generated it
Update_day sources	Explicit knowledge is not updated when it is needed, so when it is required, it may not be available
Different terminology	People, when communicating, may not understand the terminology either because they are from a different profession or another reason
Missing communication	Lack of communication for different reasons can be social, limited time, or space
Mistrust_group	Lack of companionship between the workgroup and fear to communicate the learning obtained from mistakes
Culture documentation	The company does not have a culture of writing events
Mistrust_ areas	Lack of companionship between areas of the company and fear to communicate the learning obtained from mistakes
Envy	Envy for people who know more than others so when they have new learning they do not want to share it
Fear_replacement	Fear that if a person has no "knowledge," he will become someone dispensable
Demotivation	Demotivation to document what was learned

A Solution of the Contextualized Problem: In this step, the diagnosis of each problem is made, and a result of the level of impact on the company is generated for each issue. The expert shows the current situation and offers the option that significantly impacts on the overall problem. The above is done using a machine learning technique that allows estimating the impact of the solution.

5 Application of the Proposed Method

Problem Identification and Baseline for a Knowledge Management Process:
According to the literature, a set of variables related to problems in the knowledge transfer was identified (knowledge leakage and difficulties in the coordination of teams). Results are shown in Table 1 with their interpretation. Subsequently, some software companies located in the city of Cali (Colombia) were interviewed to validate these variables. We used the Likert scale where 1 is a low score, and five is high, 0 is if it does not apply. For each question, it was asked about the impact of the problem and its frequency. Some items were added such as age, position for future studies and a question about whether or not people believe that the company has formal knowledge management (Yes, No and Don't know).

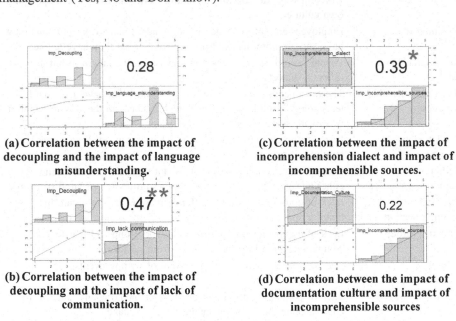

(a) Correlation between the impact of decoupling and the impact of language misunderstanding.

(b) Correlation between the impact of decoupling and the impact of lack of communication.

(c) Correlation between the impact of incomprehension dialect and impact of incomprehensible sources.

(d) Correlation between the impact of documentation culture and impact of incomprehensible sources

Fig. 2. Correlation graphs between model variables

Execution of the Plan: The characterization of this experimentation has the following data: the number of companies surveyed was 33, one survey per company was done with 16 questions and 2 answers, one of impact level and another of frequency level. Each question is related to the variables described in Table 1, which are the "x" variables of the algorithm and the variable "y" is the result of each company on whether or not knowledge management is carried out.

In the results, it was verified that there is a correlation between the variables. For example, Fig. 2a shows the existing correlation between the *disengagement impacts of the employees of the workgroup* and the *lack of understanding of the language* (The mean of the two variables is on the right).

Figure 2a shows a positive correlation (0.28) and a directly proportional relation. The fourth bar of *incomprehension in the language* is high, and in the unlinking, the last two bars are higher, thus, the greater the impact of decoupling, the greater the misunderstanding of language.

In Fig. 2b, between the *disengagement of employees* and *lack of communication*, a higher correlation is obtained. Figure 2c shows a positive relationship between *incomprehension of the dialect* and *incomprehensible sources of knowledge* (correlation coefficient of 0.39). On the other hand, Fig. 2d shows that *documentation culture* and *incomprehensible sources* had a lower correlation; this invites to analyze more thoroughly since the experts express the relationship between these variables.

A Solution of the Contextualized Problem (Machine Learning): In this stage, the variable used was associated with the question: "the company has formal knowledge management" (Yes, No and Don't know). For this variable, SVM was used to classify if these variables were leaks of knowledge and problems of team coordination. We compared three types of kernel to recognize the error and analyze it between each one. Each one of these applications of the algorithm exposes two graphs, where the first uses the training data, and the second uses the test data.

Below are the graphs that relate the impact variables of *employee untying* with *language misunderstanding*. Their interpretation is exemplified by Fig. 3 and applies to Figs. 4 and 5. According to the qualifications, it is classified that for the *untying of employees* between 2 and 4 and *incomprehension of the language* from 2 to 5, the company must have formal knowledge management. Similarly, with the *untying of employees between* 4 and 5 and a *lack of understanding of language* higher than 1.8, it is obtained that they do not formally manage knowledge. Finally, for 1–1.8 values of language compression and 1 to 4 in the separation of employees, a lack of knowledge is obtained, and a more profound study must be carried out.

In Fig. 3, for a linear kernel with a training error of 23.53%, which is considered low for this study and the test error of 56.25%, this may be due to the low amount of data used in the study. In Fig. 4, for a radial kernel with a training error of 11.76%, and a test error of 37.5%, reducing the error compared to the previous one.

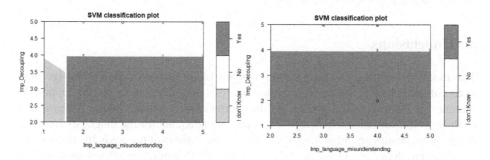

Fig. 3. Classification of train and test graphs using SVM with kernel linear.

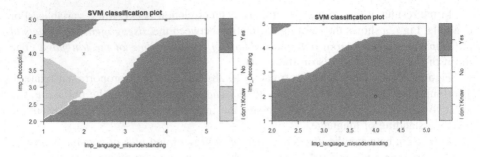

Fig. 4. Classification of train and test graphs using SVM with kernel Radial

In Fig. 5, for a polynomial kernel with a training error of 17.65%, which is considered low for this study and the test error of 25% reducing the test error what is convenient for the study.

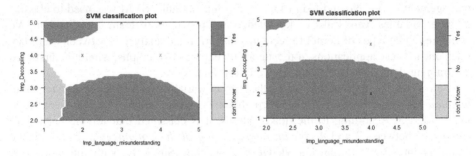

Fig. 5. Classification of train and test graphs using SVM with kernel Polynomial.

From the above, it is concluded that the variables selected by the expert to evaluate the impact of knowledge transfer in the activities of knowledge leakage and team coordination are appropriate since the SVM algorithm shows that the values of Test and entry error for different kernel types are close and allowed for this study. On the other hand, the correlation between the variables of incomprehensible knowledge sources and documentation culture should be deepened if, for the expert, the relationship is strongly related by experience or not.

6 Conclusions

The machine learning can support the analysis of the problems in knowledge management since it is based on information analysis and generate results of support for searches of solutions. The graphs and the correlation coefficient make it possible to evaluate if the variables selected by the expert are adequate for the diagnosis since this information creates an initial idea of whether in the context of the study the relationship of these problems is presented. Likewise, the SVM algorithm allows generating a classification of the problems obtaining a prediction of whether said problems affect the condition of having formal knowledge management and presents the percentage of error of the results for both the test and the training. The above generates reliability using the same knowledge management that the algorithm will improve as it adjusts.

It is evidenced that the selection of variables based on bibliographic review generates essential support as well as the precision of a diagnosis related to activities of knowledge transformation. This method is possible to apply to a broader context and different topics since it is based on steps and use of an algorithm in the same way applicable according to the need of the expert to make a diagnosis with study variables where there is a variable that depends on the others to allow classification. The precision offered by the use of automatic techniques such as (SVM) to analyze different variables and scenarios, helps to improve the certainty of concepts issued by experts in knowledge management versus how to address knowledge management issues and in particular knowledge transfer.

Finally, it is concluded that the method presented in this article has the facility to collect information in a more standardized manner, which allows a faster analysis, less wear and with greater precision in the selection of study variables for the diagnosis. It also enables the consolidation of information to contextualize the problems generating a greater focus on the issues and reducing the working time of the expert.

It is expected that in the future, recommendations can be supported by machine learning and that step 2, in which the expert must generate the filtering of the data, is also carried out by a program based on the feedback of the historical information that feeds the method.

Likewise, it is expected this method can be applied as a control system for knowledge management, performing monitoring that generates results when the incidence of each problem changes in the company thus suggesting corrective actions on knowledge leaks.

References

1. Capilla, R., Jansen, A., Tang, A., Avgeriou, P., Babar, M.A.: 10 years of software architecture knowledge management: Practice and future. J. Syst. Softw. **116**, 191–205 (2016)
2. Caiza, J.C., Guamán, D.S., López, G.R.: Herramientas de desarrollo con soporte colaborativo en Ingeniería de Software. Scielo **6**, 102–116 (2015)
3. Rashid, M., Clarke, P.M., O'Connor, R.V.: A systematic examination of knowledge loss in open source software projects. Int. J. Inf. Manag. **46**(December 2017), 104–123 (2019)
4. Wang, Y., Huang, Q., Davison, R.M., Yang, F.: Effect of transactive memory systems on team performance mediated by knowledge transfer. Int. J. Inf. Manag. **41**(April), 65–79 (2018)
5. Wang, X., Wang, J., Zhang, R.: The optimal feasible knowledge transfer path in a knowledge creation driven team. Data Knowl. Eng. **119**(January), 105–122 (2019)
6. Santoro, G., Vrontis, D., Thrassou, A., Dezi, L.: The Internet of Things: building a knowledge management system for open innovation and knowledge management capacity. Technol. Forecast. Soc. Change **136**, 347–354 (2018)

7. Zhang, X., Long, C., Yanbo, W., Tang, G.: The impact of employees' relationships on tacit knowledge sharing. Chin. Manag. Stud. **9**, 611–625 (2018)
8. Patalas-Maliszewska, J., Kłos, S.: Knowledge network for the development of software projects (KnowNetSoft). IFAC-PapersOnLine **51**(11), 776–781 (2018)
9. Nonaka, I., Takeuchi, H.: How Japanese Companies Create the Dynamics of Innovations. Oxford University Press, New York (1995)
10. Muñoz, J.L.J.: MARCO DE TRABAJO COLABORATIVO PARA APOYAR LA GESTIÓN DE CONOCIMIENTO, DESDE UN ENFOQUE DE GAMIFICACIÓN, PARA MICRO Y MEDIANAS EMPRESAS DEL SECTOR DE TECNOLOGÍAS DE LA INFORMACIÓN (2017)
11. Argote, L., Ingram, P.: Knowledge transfer: a basis for competitive advantage in firms. Organ. Behav. Hum. Decis. Process. **82**(1), 150–169 (2000)
12. Fehér, P., Gábor, A.: The role of knowledge management supporters in software development companies. Wiley Intersci. (June 2008), pp. 251–260 (2006)
13. Nonaka, I., Von Krogh, G., Voelpel, S.: Organizational knowledge creation theory: Evolutionary paths and future advances. Organ. Stud. **27**(8), 1179–1208 (2006)
14. Géron, A.: "The Machine Learning Landscape", in Hands-On Machine Learning with Scikit-Learn and TensorFlow: Concepts Tools and Techniques to Build Intelligent Systems, p. 4. O'Reilly Media Inc., California (2017)
15. Murphy, K.P.: Machine Learning a Probabilistic Perspective. MIT Press, Massachusets (2012)
16. Cortes, C., Vladimir, V.: Support-vector networks. Springer **20**, 273 (1995)
17. Yu, L., Wang, S., Lai, K.K., Zhou, L.: Bio-inspired credit risk analysis: computational intelligence with support vector machines. Springer (2008)
18. Studio, I.W.: SVM node (2019). https://dataplatform.cloud.ibm.com/docs/content/wsd/nodes/svm.html. Accessed 02 Aug 2019
19. Alshayeb, M., et al.: Challenges of project management in global software development: a client-vendor analysis. Inf. Softw. Technol. **80**, 1–19 (2016)
20. Martínez, D., Milla, A.: Mapas estratégicos. Díaz de Sa, Madrid (2012)
21. Cegarra, J.G., Martínez, A.: Gestión del conocimiento. Una ventaja competitiva, Madrid (2018)
22. Jurado, J.L., Garces, D.F., Merchan, L., Segovia, E.R., Alavarez, F.J.: Model for the improvement of knowledge management. Springer, pp. 142–151 (2018)

Linked Data and Musical Information to Improvement the Cultural and Heritage Knowledge Management

Nelson Piedra$^{(\boxtimes)}$ and Jean Paúl Mosquera Arévalo

Universidad Técnica Particular de Loja, Loja, Ecuador
{nopiedra, jpmosqueral}@utpl.edu.ec

Abstract. Knowledge management plays a crucial role in initiatives that promotes the discovery and sharing of cultural and heritage informational resources from diverse and autonomous organizations and initiatives. Information in this domain is heterogeneous, distributed, multi-lingual, comes in unstructured formats and large quantities, is strongly contextualized by time and place is created collaboratively. These challenges can be undertaken using semantic technologies. Semantic Web approach and Linked Data technologies are a catalyst for cross-domain and cross-organizational semantic data interoperability and data integration. In this regard, this work will look at the interaction between knowledge management and linked open data within the context of cultural and heritage data with a particular focus on the concept of music data. This work will discuss, among other things, how Semantic Knowledge Graphs based on linked data can contribute to enhancing knowledge management and increase productivity, in actions related with cultural data enrichment, data exploration, knowledge discovery, cultural data reuse, and data visualization. By representing information using data models and open semantic standards, data integration and reasoning can be applied to the cultural data in a well-defined way.

Keywords: Linked open data · Knowledge graphs · Music · Ontologies · Cultural heritage

1 Introduction

Classically Information related to culture, such as historical events, personalities, literary works, pictorial works, musical works, seek to be preserved, and there are contained within computerized records. Relevant information stored in these data sources exists either as structured data or as free-text. The structured data may be organized in a proprietary way or be coded using one-of- many coding, classification or terminologies that have often evolved in isolation and designed to meet the needs of the context that they have been developed. Unstructured data is content that either does not have a pre-defined data model or is not organized in a pre-defined manner. Unstructured information is typically text-heavy, but may contain data such as personal information, dates, numbers, heterogeneous entity collections, and facts as well.

Generally, these records are single class based though some records are case based and refer to multiple members of a collection. Some more informal record systems

© Springer Nature Switzerland AG 2020
J. Mejia et al. (Eds.): CIMPS 2019, AISC 1071, pp. 141–155, 2020.
https://doi.org/10.1007/978-3-030-33547-2_12

(e.g. cultural digital books) may be shared and accessed by more people. Although they are complex and heterogeneous, their scope is much more limited than the emerging use of structured data in cultural and heritage knowledge management systems.

The exponential growth of information has led to the use of Linked Open Data paradigm (LOD) for the management, representation, sharing, and linking of data in the Web [1]. This paradigm is being used in various contexts, for example, in museums, libraries, government, environment, and art, among others. In the last time various initiatives-as the proposed by the Library of Congress (LOC), Library National of France (BNF in French), and others-, are clear evidence of the interest world's interest in the LOD paradigm and semantic technologies [2].

Tim Berners-Lee [3] explains the essential facts about this paradigm and mentions that the basis of linked data is in making links into the data. Besides, he states that people or machines can search and explore the web of data, providing a broad context and meaning about this data. Since the propose of the LOD, the use of this continues to increase [4]. Ontologies are commonly used by the Semantic Web to define the concepts and relationships employed to describe an represent an area of concern allowing the classification of terms, characterization of relationships and definition constraints enabling the process of linking of data [5].

The NoSQL approaches are important currently because they allow the management of massive, heterogeneous and dynamic data volumes, for this reason, they are currently very employed by the industry, in addition to allowing the management of highly connected data and complex queries. These approaches can adapt their structure with the Linked data design guides to develop a Knowledge Graph-based in semantic model and linking this data [6, 7].

Cultural data have become complex and larger datasets, and better linking and processing has scope to transform the way we manage Cultural information. Achieving interoperability among these cultural systems continues to be a major challenge in cultural context. We have proposed that linked data and knowledge graph approaches help understand the semantics of different cultural data. The use of entity relationship models for representation of cultural data is probably not a sustainable approach. This work proposes a systematic approach for using semantic knowledge graphs and ontologies to maximize the potential of semantic interoperability when working with complex cultural datasets. Different museums, libraries, and cultural information projects are adopting linked Data technologies. Linked Data has potential to interrelate and make sense of large amounts of diverse social and cultural data as well as social or academic data.

Managing the diversity of data perhaps provides both the largest opportunities and largest challenges, especially in digital preservation and reuse of Cultural and Heritage Knowledge. Improvements in semantic interoperability of cultural and heritage knowledge management systems could contribute to the evolution from being tightly controlled local and isolated systems, to connected, interoperable and complex systems built upon heterogeneous and linked data sources. In the first part of this work, related works are described. In the second part, the authors propose a cycle to transform existing information into linked data formats. The third part focuses on the extract of cultural data from non-structured data sources. In the fourth part, the authors focus on musical context, ontology to represent musical data is created, and data extracted are cleaned, mapped to ontology and transformed to linked data. Then, the data generated will be enriched with knowledge pieces from the Linked Open Data cloud. Finally, the data are published and a SPARQL endpoint is enabled to ad-hoc queries. The result is that the dataset is more visible on the open web, and brief recommendations are made.

2 Related Works

Heritage itself is conceptualized as the meanings attached in the present have been passed down from previous generations and is regarded as a knowledge defined within history, social, and cultural contexts. The preservation and dissemination of information of a country–such as their historical and cultural characteristics– have been of interest in the last years. Efforts have focused on digitizing, storing, and managing this information. The principal purpose is the digital preservation, openness, reuse and dissemination of these essential components of a nation's identity.

The very first project to digitize cultural texts and make them freely available was Project Gutenberg, which started in 1970. Today there are several organizations and projects to unify historical and cultural to allow access to these, among others, the digitized content Digital Public Library of America (DPLA, that supports direct search, view by Timeline, Map view, and Thematic Exhibitions); HathiTrust; Digital Collections at the Library of Congress and Internet Archive (see: https://archive.org/); in Europe, several projects focused on historical and cultural preservation have been developed, the most important is Europeana (over 55 million "artworks, artifacts, books, videos, and sounds from across Europe", see collections: http://www.europeana. eu/portal/en). Both DPLA and Europeana support developers to create experimental interfaces and external applications that expands functionality to that their artifacts can be viewed and used.

Europeana datasets use the EDM model or Europeana Data Model, which consists of a model developed in this same project to integrate, collect, and to enrich the information related to the cultural heritage. In general, the content provider's shares data using common vocabularies: OAI ORE (Open Archives Initiative Object Reuse and Exchange), SKOS (Simple Knowledge Organization System), CIDOC-CRM. Today, thanks to this initiative, more than 50 million records related to historical and cultural information are in the RDF (Resource Description Framework) format [7, 12].

Another initiative funded by the European Commission is SmartMuseum, which consists of the development of a recommender system, to provide suggestions aimed at tourists focused on cultural heritage. In the present case, semantic technologies are used for the treatment of heterogeneous information, which comes from content descriptors, sensors, and user profiles. SmartMuseum uses the General Ontology User Model Ontology (GUMO), in conjunction with the vocabularies: W3C Geo Vocabulary for Geographic Information modeling, Getty Vocabularies for object indexing. During the year 2011, the British Museum had 2 million records of its database in Linked Open Data. Currently, more than 100 million records are linked under the model CIDOC-CRM [12].

Damova and Dannells [18] raise methods for the creation of a semantic knowledge base of museums information and to relate it with Linked Open Data (LOD). The models uses Proton, an ontology containing 542 classes of entities and 183 properties [19]. It also uses the model CIDOC-CRM, which is an ontology designed by the International Council of Museums to provide a scheme to address this area of knowledge, is composed of 90 classes and 148 properties [20]. This project focuses on the treatment of information from the Museum of the city of Gothenburg. A study related to the publication and use of linked data related to the cultural heritage is described in [21], where it performs an inductive analysis initiating with a proposal of a business model of a portal on cultural inheritance. Which it is aligned with the requirements to publish linked data and culminated with a proposal of an environment of search and recommendation of this information.

In [1] they propose a method to describe the catalogs of musical compositions and events of the National Library of Paris, Radio France and the Philharmonic Orchestra of Paris semantically. Among the models used are CIDOC-CRM and FRBRoo, managing to present the knowledge network called DOREMUS, which comprises three linked data bank (linked datasets) that describe works of classical music and events associated with these like concerts, among others. A transparent and interactive methodology for the consumption, conversion, and link of cultural information in Linked Data is proposed in [22]. Where they use the conversion tool XML-RDF contained in the framework ClioPatria, achieving convert the metadata from the Museum of Amsterdam Linked data under the model European Data Model (EDM). In [23] They propose an ontology for the organization, classification, and mining of documents related to the culture. The work describes a human-machine system that uses statistical methods of natural language processing for the extraction of cultural information from texts.

Knowledge Graphs are an important application to the LOD and semantic technologies using them makes it possible to develop recommendation systems, predictions, fraud detection systems, and optimize the discovering and searching of data. Google is an example of this application: The Google Knowledge Graph provides people summarized information about a search, for example, if the search is based on a famous painter, Google shows linked information about their works, similar painters and other details [8–10].

In [11], authors propose the use of a NoSQL scheme based on graphs for the treatment of information about historical figures of Ecuador, for this use Neo4j as a database engine for the management and visualization of this information. This work is an approach to the management of heterogeneous data. In [13] Benjamins et Al. describe a case study related to ways of using Semantic Web technologies for the treatment of information related to cultural heritage, which characterizes an ontology to describe humanities, a semi-automatic tool for make annotations, also a procedure to generate a cultural portal of a semantic character. On the other hand [14] analyzes how ontologies and technology would help in the access to the information of the museums by analyzing the information of museums and concepts of the semantic web such as ontologies, schematics, among others. Signore proposes a method to ease the cataloging of artifacts in 3D related to the cultural inheritance [15], in which it addresses several Greek relics and proposes a model for the annotation of this type of objects [16]. This approach is similar to the proposal in [17] where an ontological model is proposed to address archaeological information, specifically resources in 3D.

3 Proposed Strategy

Currently, various authors widely debate the approach to discovering and reuse information sources containing cultural data. The purpose is to provide open access to a vast amount of cultural resources hosted by several institutions such as libraries, archives, museums, galleries, and audiovisual collections. It is indeed an ambitious goal that needs a widely consensus-based approach for the implementation of political and technological agreements among the involved institutions. Information in this domain is heterogeneous, distributed, multi-lingual, comes in unstructured formats and large quantities, is strongly contextualized by time and place is created collaboratively. These challenges can be undertaken using semantic web technologies.

Semantic Web (SW) approach and Linked Data (LD) technologies are a catalyst for cross-domain and cross-organizational semantic interoperability and data integration. Several authors describe LD as a term used to describe best practices for sharing, linking data and knowledge in the Semantic Web [34, 35]. Any Linked Data approach can be expected to have three phases, presented in Fig. 1: data and metadata acquisition, Linked Data production and data exploits interface. The essential objective of this approach is to connect related data that was not previously linked:

First, the data acquisition phase is in charge of retrieving data from data sources. In cases where the retrieved data is not in the RDF format, wrappers can be used to extract the content into structured data. Several tools are available that can be used when implementing a data access component. The proposal implements a wrapper module to access heterogeneous data sources in which wrappers are built around each data source to provide a unified view of the retrieved data. Some websites provide APIs to access the data; SPARQL Client Libraries provide a way for accessing SPARQL endpoints.

Second, following the approach proposed in [24], the LD transformer phase is responsible for ontology modeling, cleaning and disambiguate, mapping, and linking the extracted data to produce new Linked Data. This phase is executed through a cycle of continuous improvement described below. This cycle consists of five steps, business and data understanding, choosing the right and clear path of modeling such data can have many advantages. Data extraction and cleaning, these tasks are necessary to working with scattered and unstructured data; Ontological modeling; Transformation and linking; and, Continuous improvement. Our objective is to use this process to extract, disambiguate, transform and visualize cultural knowledge from heterogeneous data sources as LD. However, we concentrated through the prism of the Ecuadorian music context. Choosing the right path of modeling such data can have many advantages.

Third, the User interface phase provides a way of interacting with the application. This will often, but not necessarily, be a Web interface. The data exploits may include a user interface supporting, for example, Sparql ad-hoc queries of the RDF data and also an API for programmable interaction with the system. By sharing meanings and data using data models and open standards, such as RDF, OWL, SKOS, and, data integration and reasoning can be applied to the data in a well-defined way. The proposed architecture is shown below.

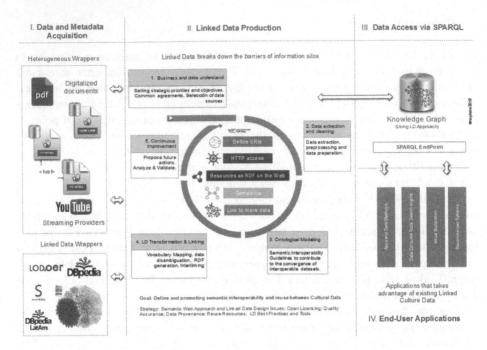

Fig. 1. General strategy to LD production and data exploitation

4 The Phase of Data Sources Discovering

This section describes the process of data source selection, data extraction and synthesis for information related to Ecuadorian music. One of the main phases is the data sources discovering, this project is focused on data from the Internet (blogs, video channels, and others.), historical books, and related works with a close relationship with the popular music of Ecuador. The disadvantage of this sources is the heterogeneous nature of this sources related with the structure, format, and type of source, and this leads the use of a schema NoSQL to manage this data as detailed in [25] by Wu et al.

Table 1. Sources of information

Source	Description or characteristic	Examples
Books related historical music of Ecuador	This source contains history about the music of Ecuador	Bibliografía de la Música Ecuatoriana [26]
Websites about the historical music of Ecuador	Blogs, Youtube Channels related to Popular Music and traditional genres of Ecuador	Blog del pasillo [27]
Videos about popular songs	Videos related Popular Music of Ecuador	Musicoteca Ecuador [28]
Another sources	Diversity of sources related to Ecuadorian Music	Lo mejor del Siglo XX: Musica Ecuatoriana [29]

This information sources (see Table 1) talk about the traditional Ecuadorian music, for example, data related to the songs, artist information, history, and other information related to the music. Some Traditional genres are described in Table 2, these genres are part of the musical heritage of Ecuador.

Table 2. Popular traditional genres of Ecuador

Genre	Description
Pasillo Ecuatoriano	Traditional Genre of Ecuador, this denomination is because their planimetric structure is of often steps. Within the artist for this genre is, for example, Salvador Bustamante Celi. Julio Jaramillo, Carlota Jaramillo, and others [30]
Pasacalle	Traditional Genre of Ecuador characterized by their rhythm that is cheerful is of Spanish origin. Within the artist for this genre are Alfredo Carpio, Rafael Carpio Abad, and others [31]
Villancico	A traditional genre of Ecuador, listen in Christmas Festivities, within the artist for this genre are, for example, Salvador Bustamante Celi, Guillermo Garzon, and others
Other genres	Sanjuanito, Tonada, Yaravi, and others

Within the information related to the artist, it is important to mention that an artist can be a composer, performer, and Author; Table 3 describes their characteristics.

Table 3. Characterization of artist entity

Genre	Description
Artist	Is a person that has participated in a musical work
Composer	Is an artist that compose the music of a song or a musical work
Author	Is an artist that writes the letter of a song or a musical work
Performer	Is an artist that sings or interpret a song or a musical work

Figure 2 shows the existing relationship into a Musical work with an author, composer, and Performer.

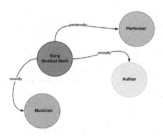

Fig. 2. Relationships between concepts: musical work, composer, author and performer.

5   The Phase of Data Extraction and Cleaning

Once established the sources and categories of data, this work focuses on obtaining the information of these sources. The complex structure of the sources leads to the use of a manual process, besides its nature (i.e. unstructured contents). Several novelties were presented, among which are, metadata incomplete, incomplete and non-hardened information.

The data sources analyzed presented some heterogeneity among them in relation to the metadata and structure of these, this led to the individual search of specific compositions in other sources to obtain as much information as possible about it, in lack of an official source or complicated access to it, i.e., its access is not public.

The result of the manual extractions was 1115 records of the information available on the sources analyzed. The next step consisted of the process to clean the data (e.g., remove duplicated records, normalize the matrix). The result of all this process was the setting of 600 artists within people and groups, 18 musical genres, 8 countries, and nationalities. Technicians looking to work and collaborate with large cultural datasets and looking to work with data need to extend their capability to deliver semantically interoperable systems by arming themselves with an ontological approach that will boost adoption of ontological and non-ontological resources, and encourage participation of domain experts.

6   The Phase of Semantic Representation: Ontological Modeling

There is a clear trend of adopting linked open data and ontologies for enabling semantic interoperability by key stakeholders that facilitate cultural and heritage information exchange. The most important value offered by the presented approach in this context is the ability to allow autonomous systems of communicating the meaning of similar concepts used within the domain.

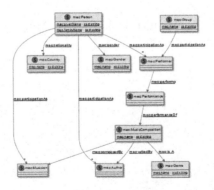

Fig. 3. Mec ontology

For this phase different ontologies were analyzed, the search of vocabularies led to discovering of potential vocabularies for the entities described previously, within the ontologies and models analyzed are Music Ontology, DBpedia, GeoNames, Schema, Getty, European Union Vocabularies and other following detailed. Music Ontology (mo) provides a model for publishing structured music-related data on the web of data. The ontology contains 54 classes and 153 properties (e.g. Artist, Musical Artist, Group Artist, Tracks, and Album. This ontology is maintained by Yves Raimond et Al. [32]. Another ontology analyzed is DBpedia (dbo) a cross-domain ontology that has been manually created based on the most commonly used infoboxes within Wikipedia. The DBO ontology currently covers 685 classes which form a subsumption hierarchy and are described by 2,795 different properties [33]. Other ontologies and vocabularies studied were Wikidata[1], Schema[2], Getty Ontology[3].

As we move towards achieving more comprehensive levels of semantic interoperability, linked open data and ontologies have proved to be more dynamic than other methods used. The Table 4 describes the Ecuadorian Music Ontology (mec Ontology) this ontology was created to describe a musical composition. This ontology also allows defining information related to the creators and performers of the song (i.e., name and last name, nationality, gender). Define general aspects about a song, or about the people that were part of the musical creation (authors, composers, performers). The ontology proposed can be expanded with more attributes of a musical work or music composition, being possible describe a song with more details, for example, talk about the musical instruments, lyrics, related videos, and other awareness information. See Fig. 3 for a partial view of the proposed ontology.

Nevertheless, there are some limitations of ontologies as they are scaled to represent information from large complex information domains. Building large ontologies can be resources consuming and can require considerable amount of input from domain experts. These methodological issues need to be resolved in order to have mainstream adoption that will demonstrate the collective effect of using ontologies for better semantic interoperability and data integration within the cultural and heritage ecosystem.

7 LD Transformation, Linking and Sparql Queries

Initiatives related to culture and heritage reflects huge data, but it is difficult for individuals to access and discover this information, so a well-chosen methodology is crucial. The basic measure for mining such data is transforming this into an understood

---

[1] According [36] "is a free, collaborative, multilingual, secondary database, collecting structured data to provide support for Wikipedia, Wikimedia Commons, the other wikis of the Wikimedia movement, and to anyone in the world".

[2] For [37] "Schema.org is a collaborative, community activity with a mission to create, maintain, and promote schemas for structured data on the Internet, on web pages, in email messages, and beyond".

[3] Getty Research [38] says "The Getty vocabularies contain structured terminology for art, architecture, decorative arts, archival materials, visual surrogates, conservation, and bibliographic materials".

Table 4. Reused vocabularies

Class	Property	Domain class	Range
Author, Compositor, Performer	foaf:givenName	foaf:Person	xsd:string
	foaf:familyName	foaf:Person	xsd:string
	schema:gender	foaf:Person	schema: GenderType
	schema:nationality	foaf:Person	schema:Country
	mec:participateAs	foaf:Person	mec:Author
	mec:participateAs	foaf:Person	mec:Musician
	mec:participateAs	foaf:Person	mec:Performer
Musical group	foaf:name	foaf:Group	xsd:string
	mec:participateAs	foaf:Group	mec:Performer
Genre	dbo:label	dbo:Genre	xsd:string
Gender	schema:gender	schema: GenderType	xsd:string
Music composition	mec:title	mec: MusicComposition	xsd:string
	mec:writedBy	mec: MusicComposition	mec:Author
	mec:composedBy	mec: MusicComposition	mec:Musician
	mec:is_a mec:verse	mec: MusicComposition	dbo:Genre
Performance	mec: performanceOf	mec:Performance	mec: MusicComposition
	mec:performs	mec:Performer	mec:Performance

structure. The next phase was the transformation of information using the scheme according to the guidelines proposed by Piedra et al. [24, 34]. In Fig. 4 is described a very popular song in Ecuador: "Vasija de Barro," of genre Danzante, was written by the authors Jaime Valencia, Hugo Aleman, Jorge Adoum and Jaime Valencia, and composed by Gonzalo Benitez and Luis Alberto Valencia.

Using proprietary tools made for conversion, the data was converted in XML/RDF format. Finished the process, the RDF generated was validated using the service RDF Validator from W3C (https://www.w3.org/RDF/Validator/rdfval). The Linked Data generated was linked to external data sources.

Linking techniques was also applied, since it may allow instances to link to external data resources (e.g. Dbpedia, Wikidata, LOC, BNE, BNF) as a precursor to building an Ecuadorian Cultural and Heritage RDF triple store (with a SPARQL endpoint). The use of owl: sameAs further enriches the Linked Data space by declaratively supporting distributed semantic data integration at the instance level. In this stage, we conduct an pilot of linking, see Table 5.

Table 5. Extract of linked with external data-sources

Class	Generated resources (mec)	dbo	wikidata	loc	bne*	bnf*
Genre	mec:Huayno	http://dbpedia.org/page/Huayno	https://www.wikidata.org/wiki/Q2307786	http://id.loc.gov/authorities/genreForms/gf2015026022		
Genre	mec:Mapalé	http://dbpedia.org/page/Mapal%C3%A9	https://www.wikidata.org/wiki/Q1423598			
Genre	mec:Pasillo	http://dbpedia.org/page/Pasillo;	https://www.wikidata.org/wiki/Q7142141			
Genre	mec:Pasodoble	http://dbpedia.org/page/Pasodoble	https://www.wikidata.org/wiki/Q208341	http://id.loc.gov/authorities/subjects/sh85098443		
Genre	mec:Salsa	http://dbpedia.org/page/Salsa_music	https://www.wikidata.org/wiki/Q208239	http://id.loc.gov/authorities/genreForms/gf2014027066		
Genre	mec:Sanjuanito	http://dbpedia.org/page/Sanjuanito	https://www.wikidata.org/wiki/Q6120148			
Genre	mec:Villancico	http://dbpedia.org/page/Villancico	https://www.wikidata.org/wiki/Q503354	http://id.loc.gov/authorities/subjects/sh85143362		
Genre	mec:Yaravi	http://dbpedia.org/page/Yarav%C3%AD		http://id.loc.gov/authorities/names/no98044726		
Person	mec:AgustinCueva	http://dbpedia.org/page/Agustin_Cueva	https://www.wikidata.org/wiki/Q382260	https://id.loc.gov/authorities/n79110733	http://datos.bne.es/resource/XX1133619	

We have placed strong emphasis in a decoupled approach, where the components of the architecture maintain autonomy. Our final goal is to contribute in making available more semantic cultural information while keeping a lower entry barrier for developers.

The creation of a global data-space of easily reusable, and semantically rich metadata elements would enable the cultural systems to concentrate increasingly disperse resources on adding unique value. Participation in the development of such global data-space should also enable the improved integration and visibility of organization and begin to take the possibilities indicated by the current linked open data cloud to new levels of utility.

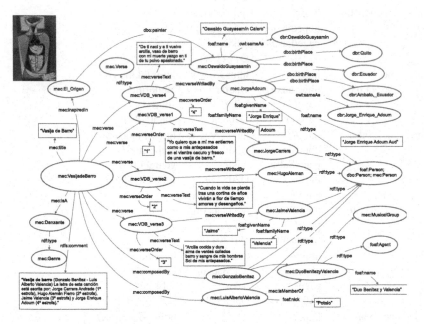

Fig. 4. Graph representation of the song "Vasija de Barro."

Queries for getting specific information such as to filter the data for obtaining all the songs by a specific genre, all the song written by a specific Author, or all the songs musicalized by a specific musician. Below is a search to filter the songs written by a particular author.

```
PREFIX mec: < http://ld.utpl.edu.ec/schema/mec#
SELECT ?Person WHERE{
    ?Person mec:nationality mec:Ecuador.}
```

The last query returns all the artists from Ecuador. The result of this query is 433 artists of this country. This is an example of the possible queries that could be done with this data. This initiative seeks to continue and encompass more fields of art like the literature, painting, and other cultural expressions from Ecuador. The following query is to filter the musical compositions from their genre, specifically founds all the song with are part of the genre "Pasillo."

```
PREFIX mec: < http://ld.utpl.edu.ec/schema/mec#
SELECT ?MusicComposition WHERE{
     ?MusicCompositon mecs:genre mec:Pasillo .}
```

The result of this query returns all the songs that are a "Pasillo" (Popular Genre of Ecuador), about 409 songs correspond to this genre. Another possible query is filtering the artist by their gender.

```
  PREFIX mec: < http://ld.utpl.edu.ec/schema/mec#
  SELECT ?Person WHERE{
?Person mecs:gender mec:Female .}
```

The last query returns all the artists that are Woman, in the consolidated Knowledge Base 46 artists are Woman, and 406 artists have the Gender Male. To fill the gaps in the generated collection and to make it more interesting and useful for all users, authors propose to listen to end-users and elicit their needs for cultural heritage content to prioritize content acquisition, transformation, linked and publication. User demands are also important to understand in relationship with discoverability, usability, and reusability of cultural data.

8   Conclusions

This work will look at the interaction between cultural knowledge management and linked data within the context of Ecuadorian musical data. Linked data can be considered as a representation of the current technological capability to represent and connect a huge amount of data through heterogeneous, distributed and autonomous data sources that can then be analyzed to provide valuable information and knowledge.

The proposed cycle transforms existing musical information into linked data formats and publishes it to the RDF triple store. The data generated is enriched with knowledge pieces from the Linked Open Data cloud. The end result is that the dataset is more visible on the open web. Improving openness and data and semantic quality is the number one priority in achieving better access to cultural heritage. Improving the structuration and quality of existing data while actively searching for new data from new cultural data sources to be added to the knowledge graph is key to content development.

Once the RDF data is available in a triple store, it can be used and accessed through the logic and presentation layers. Some of the logic layers may be implemented in the data by reasoning over the triple store. As future work, other forms of processing that cannot be implemented on the data may be carried out with the application of statistical or machine learning processes to make inferences from the data. Additionally, as future work, it is planned to extend the query functionality towards a federated SPARQL engine. Federated SPARQL Engines provide a single access point for multiple heterogeneous data sources. There is a lot more content in museums, libraries, archives, and audiovisual institutions than we can link and visualize in the context of this work in a meaningful way.

Acknowledgement. The work has been funded and supported by the Universidad Técnica Particular de Loja (UTPL), through the KBS Research Group.

References

1. Achichi, M., Lisena, P., Todorov, K., Delahousse, J.: DOREMUS : a graph of linked musical works. In: International Semantic Web Conference, vol. 1, pp. 3–19 (2018)
2. Vattuone, M.M.S., Roldán, J., Neder, L., Maldonado, M.G., Vattuone, M.A.: Formative pre-Hispanic agricultural soils in northwest Argentina. Q. Res. **75**, 36–44 (2011). https://www.sciencedirect.com/science/article/pii/S0033589410001043
3. Berners-Lee, T.: Linked Data (2006). https://www.w3.org/DesignIssues/LinkedData.html
4. Hutchison, D.: The Semantic Web – ISWC (2013)
5. W3C. Vocabularies (2018). https://www.w3.org/standards/semanticweb/ontology
6. Silvescu, A., Caragea, D., Atramentov, A.: Graph databases. Artif. Intell. **14** (2010). http://www.cse.iitk.ac.in/users/smitr/PhDResources/GraphDatabases.pdf
7. Charles, V., Isaac, A., Manguinhas, H.: Designing a multilingual knowledge graph as a service for cultural heritage – some challenges and solutions. In: Proceedings of the 2018 International Conference on Dublin Core and Metadata Applications, pp. 29–40 (2018)
8. Barker, S.: Google knowledge graph (2018). https://hackernoon.com/everything-you-need-to-know-about-google-knowledge-graph-and-how-to-get-included-e14c07f95fe6
9. Koide, S., Kato, F., Takeda, H., Ochiai, Y.: Action planning based on open knowledge graphs and lod. In: JIST (Workshops & Posters) (2017)
10. Noev, N., Bogdanova, G., Todorov T., et al.: Using graph databases to represent knowledge base in the field of cultural heritage. In: Proceedings of the Sixth UNESCO International Conference on Digital Presentation and Preservation of Cultural and Scientific Heritage (2016)
11. Mosquera, J., Piedra, N.: Use of graph database for the integration of heterogeneous data about Ecuadorian historical personages. In: 2018 7th International Conference Software Process Improvement, pp. 95–100 (2018)
12. Lodi, G., Asprino, L., Giovanni, A., Presutti, V., Gangemi, A., Reforgiato, D., et al.: Semantic web for cultural heritage valorisation. In: Data Analytics in Digital Humanities, pp. 3–37 (2017)
13. Benjamins, V., Contreras, J., Blázquez, M., Dodero, J., Garcia, A., Navas, E., et al.: Cultural heritage and the semantic web. In: European Semantic Web Symposium, pp. 433–44. Springer, Heidelberg (2004)
14. Signore, O., National, I.: The semantic web and cultural heritage : ontologies and technologies help in accessing museum information. In: Information Technology for the Virtual Museum (2015)
15. Yu, C., Groza, T., Hunter, J.: Reasoning on crowd-sourced semantic annotations to facilitate cataloguing of 3D artefacts in the cultural heritage domain. In: International Semantic Web Conference, pp. 228–243 (2013)

16. Castiglione, A., Colace, F., Moscato, V., Palmieri, F.: CHIS: a big data infrastructure to manage digital cultural items. Future Gener. Comput. Syst. **86**, 1134–1145 (2017). http://dx. doi.org/10.1016/j.future.2017.04.006
17. Ben Ellefi, M.: Cultural heritage resources profiling : ontology-based approach. In: Companion Proceedings of the Web Conference (2018)
18. Damova, M., Dannells, D.: Reason-able view of linked data for cultural heritage, 2–9 (2011)
19. Semantic Web. PROTON, a lightweight upper-level ontology. http://proton.semanticweb. org/
20. ICOM, CIDOC. Conceptual Model Reference CMR (2018). http://www.cidoc-crm.org/
21. Hyvönen, E.: Publishing and Using Cultural Heritage Linked Data on the Semantic Web. Morgan & Claypool Publishers, San Rafael (2012)
22. de Boer, V., Wielemaker, J., van Gent, J., Hildebrand, M,, Isaac, A., van Ossenbruggen, J., et al.: supporting linked data production for cultural heritage institutes: the amsterdam museum case study. In: Proceedings 9th International Conference Semantic Web Research Application, pp. 733–747. Springer, Heidelberg (2012). http://dx.doi.org/10.1007/978-3-642-30284-8_56
23. Mavrikas, E.C., Nicoloyannis, N., Kavakli, E.: Cultural heritage information on the semantic web. In: International Conference on Knowledge Engineering and Knowledge Management, pp. 3–5 (2018)
24. Alexander, S., Piedra, N., Morocho, J.C.: Using linked data to ensure that digital information about historical figures of loja remains accessible and usable (2018)
25. Wu, Q., Chen, C., Jiang, Y.: Multi-source heterogeneous Hakka culture heritage data management based on MongoDB. In: 2016 5th International Conference Agro-Geoinformatics, (Agro-Geoinformatics 2016) (2016c)
26. Guerrero, F.: Bibliografia de la Musica Ecuatoriana en linea. 0.1c. Quito: Ministerio de Cultura y patrimonio de Ecuador (2017). https://bibliografiamusicalecuatorianaenlinea. wordpress.com/
27. Herrera, C.: Pasillo Ecuador (2018). http://pasilloecuador.blogspot.com/
28. Eche, J.: Musicoteca Ecuador (2018). http://musicotecaecuador.blogspot.com/2016/09/que-es-musicoteca-ecuador.html
29. Carrión, O.: Lo mejor del siglo XX: música ecuatoriana. Duma (2002)
30. Wong, K.: La nacionalización del pasillo ecuatoriano a principios del siglo XX. III Congr Latinoam la Asoc Int para el Estud la música Pop (1999)
31. Denizeau, G.: Los géneros musicales, una visión diferente de la Historia de la música. Robinbook, editor. Barcelona (2002)
32. Raimondm, Y., Gängler, T., Giasson, F., Jacobson, K., Fazekas, G., Reinhardt, S., et al.: Music ontology (2013). http://musicontology.com
33. Dbpedia. Dbpedia Ontology (2018). https://wiki.dbpedia.org/services-resources/ontology
34. Piedra, N., Chicaiza, J., Lopez, J.: Guidelines to producing structured interoperable data from Open Access Repositories An example of integration of digital repositories of higher educational institutions LatAm (2016)
35. Linked Data Org. Linked Data (2018). http://linkeddata.org/
36. Wikidata. About Wikidata (2018). https://www.wikidata.org/wiki/Wikidata:Introduction
37. Schema Org. Schema (2018). https://schema.org/
38. Getty Research. Getty Vocabulaires as Linked Open Data (2018). http://www.getty.edu/ research/tools/vocabularies/lod/

Software Systems, Applications and Tools

Distributed System Based on Deep Learning for Vehicular Re-routing and Congestion Avoidance

Pedro Perez-Murueta[1][✉] ⓘ, Alfonso Gomez-Espinosa[1][✉] ⓘ,
Cesar Cardenas[2] ⓘ, and Miguel Gonzalez-Mendoza[3] ⓘ

[1] Tecnologico de Monterrey, Escuela de Ingeniería y Ciencias,
Av. Epigmenio Gonzalez 500, 76130 Santiago de Querétaro, Querétaro, Mexico
{pperezm, agomeze}@tec.mx
[2] Tecnologico de Monterrey, Escuela de Ingeniería y Ciencias,
Av. General Ramon Corona 2514, 45138 Zapopan, Jalisco, Mexico
[3] Tecnologico de Monterrey, Escuela de Ingeniería y Ciencias,
Av. Lago de Guadalupe KM 3.5, 52926 Atizapán de Zaragoza,
Estado de Mexico, Mexico

Abstract. The excessive growth of the population in large cities has created great demands on their transport systems. The congestion generated by public and private transport is the most important cause of air pollution, noise levels and economic losses caused by the time used in transfers, among others. Over the years, various approaches have been developed to alleviate traffic congestion. However, none of these solutions has been very effective. A better approach is to make transportation systems more efficient. To this end, Intelligent Transportation Systems (ITS) are currently being developed. One of the objectives of ITS is to detect congested areas and redirect vehicles away from them. This work proposes a predictive congestion avoidance by re-routing system that uses a mechanism based on Deep Learning that combines real-time and historical data to characterize future traffic conditions. The model uses the information obtained from the previous step to determine the zones with possible congestion and redirects the vehicles that are about to cross them. Alternative routes are generated using the Entropy-Balanced kSP algorithm (EBkSP). The results obtained from simulations in a synthetic scenario have shown that the proposal is capable of reducing the Average Travel Time (ATT) by up to 7%, benefiting a maximum of 56% of the vehicles.

Keywords: Traffic congestion detection · Traffic prediction · Congestion avoidance · Intelligent transportation systems · Deep learning

1 Introduction

Today, traffic congestion is a great challenge for all the major cities of the world. Traffic jams are a recurrent problem in urban areas, their negative effects are easy to observe increased emissions of greenhouse gases, and carbon dioxide, damage to health, increased car accidents and economic losses. According to a recent mobility report, in

© Springer Nature Switzerland AG 2020
J. Mejia et al. (Eds.): CIMPS 2019, AISC 1071, pp. 159–172, 2020.
https://doi.org/10.1007/978-3-030-33547-2_13

the US, the economic loss generated by delays in travel time and fuel consumption was estimated at $ 121 billion and is expected to reach $ 199 billion by 2020. Therefore, finding effective solutions to minimize traffic congestion has become a major problem [1].

The implementation of vehicular network standards and advances in wireless communication technologies have allowed the implementation of Intelligent Transportation Systems (ITS). One of the purposes of ITS is real-time traffic management using vehicular data collected by various sensors installed in the road infrastructure to characterize traffic and, in this way, to detect, control and minimize traffic congestion. The main challenge of this approach is to use a combination of historical data and forecast models that allow congestion and reroute vehicles without causing a new congestion elsewhere [2].

There are commercial solutions such as Waze [3] and Google Maps [4] capable of providing alternative routes from origin to destination. Nevertheless, while these systems are capable of determining the long-term status of traffic, they often behave as reactive solutions that cannot yet prevent congestion. Likewise, the suggested routes to users are based exclusively on the algorithm of the shortest route to their destination, without taking into account the impact that the re-routing of vehicles may have on future traffic conditions [5].

In this work, we present a congestion avoidance by re-routing system that uses a mechanism that combines historical data and predictions obtained from Deep Learning architecture in order to characterize future traffic conditions to make an early congestion detection. Based on this information, an algorithm for generating k-Shortest Routes provides alternatives to all those vehicles that are about to cross possible congested areas. The system uses information obtained from probe cars for the detection of non-recurrent congestion. The results obtained from simulations in synthetic scenarios have shown that the proposal is capable of reducing the Average Travel Time (ATT) by up to 7%, benefiting and 50% of the vehicles. The rest of this document is organized as follows. In Sect. 2, previous research in this field is summarized. Section 3 presents the details of the proposal. Section 4 presents the tools used in the development of the proposal and the scenarios used for the evaluation. Section 5 shows the results. Finally, Sect. 6 presents conclusions and future work.

2   Related Work

Several static and dynamic route guidance proposals have been developed to help drivers determine the best routes for their trips. Most proposals for static guides [6, 7] focus on finding the existing route of the shortest distance, or with the shortest travel time. When determining the route, the updated traffic information is never considered, this means that this type of system does not generate adequate routes to the existing traffic conditions and does not mitigate traffic congestion in the city Dynamic guidance systems use real-time traffic information to generate the best travel routes [8]. Nevertheless, most of these proposals do not alter these routes if traffic conditions worsen during the trip, not to mention that many of these proposals consider a single central system in charge of monitoring the traffic situation.

CoTEC [9] (Cooperative Traffic congestion detECtion) is a cooperative vehicle system that uses fuzzy logic to detect local traffic conditions using beacon messages received from surrounding vehicles. When detecting congestion in the area, CoTEC activates a cooperative process that shares and correlates the individual estimates made by different vehicles, characterizing of the degree of traffic congestion. Tested in large scenarios, COTEC obtained good results in the detection of congestion; yet, it does not integrate any vehicle re-routing strategy.

Araujo et al. [10] present a system called CARTIM (Identification and minimization of traffic congestion of cooperative vehicles). Based on CoTEC, also uses fuzzy logic to characterize road traffic, adding a heuristic that seeks to reduce drivers' travel time by allowing them to modify their travel routes, improving vehicle flow in the event of congestion. The authors report that CARTIM achieves a reduction of 6% to 10% in ATT.

Urban CONgestion DEtection System (UCONDES) [11] is a proposal that uses an Artificial Neural Network (ANN) designed to detect and classify existing levels of congestion in urban roads. UCONDES also adds a mechanism that suggests new routes to drivers to avoid congested areas. The ANN uses the speed and traffic density of a street as input to determine the existing level of congestion. The vehicles receive the classification obtained, as well as a new route so that the vehicle can alter its current route and avoid congested roads. The simulations carried out in a realistic Manhattan scenario showed a reduction of 9% to 26% of ATT.

De Sousa et al. [12] present SCORPION (A Solution using Cooperative Rerouting to Prevent Congestion and Improve Traffic Condition). In this proposal, all vehicles send their information to the nearest Road Side Unit (RSU), which uses the k-Nearest Neighbor algorithm to classify street congestion. After determining the traffic conditions, a collective re-routing scheme plans new routes to those vehicles that will pass through congested areas. Like UCONDES, SCORPION performed simulations using a realistic Manhattan scenario, achieving a 17% reduction in ATT.

The proposals mentioned above present some limitations, such as a limited re-routing distance, no real-time mechanism for the detection of congestion or no mechanism for re-routing with a global perspective.

Pan et al. [5] proposes a centralized system that uses real-time data obtained from vehicles to detect traffic congestion. When congestion is detected, vehicles are redirected according to two different algorithms. First, the shortest dynamic route (DSP), which redirects vehicles using shorter routes that also have the lowest travel time. The authors mention that there is the possibility of moving the congestion to another point. To solve this, they use a second algorithm, Random k-Shortest Path (RkSP), which randomly chooses a route between k shorter routes. The objective is to avoid changing the congestion from one place to another and, in this way, to balance the redirected traffic between several routes. Regardless, this proposal does not implement a real-time mechanism to detect congestion as it occurs; it is only able to detect it during the next redirection phase.

A strategy that allows improving the aforementioned solutions is to anticipate traffic congestion, for this, it is necessary to be able to predict the future state of vehicular traffic. Kumar [13] proposes a traffic prediction system that uses a Kalman filter. This proposal uses historical and real-time information when making a prediction. Despite showing good results in terms of predictions, the proposal does not include re-routing strategies in case of congestion. Another element to consider is that the test scenario is a single street segment, which does not allow to determine its efficiency in an entire vehicular network.

From the perspective of Machine Learning, traffic prediction is a spatio-temporal sequence prediction [14]. Suppose that the traffic prediction problem in a vehicular network is a dynamic system on a spatial region represented by M x N adjacency matrix of M rows and N columns, and within each cell of this matrix, there exist P measurements, which vary with time. Based on the above, it can be seen that the problem of traffic prediction differs from a single-step time series prediction, since the purpose of the former kind of prediction involves working with a sequence containing spatial and temporal structures.

In recent times, Long Short-Term Memory Networks (LSTM) have gained popularity as a tool for the prediction of spatio-temporal sequences [15]. LSTM are a special type of Recurrent Neural Networks (RNN) that have been shown to be able to learn long-term dependencies [16, 17]. LSTMS are specifically designed to avoid the problem of long-term dependency. Each cell acts as an accumulator of state information and controlled by a set of cells that determines the flow of information. There is a gateway that allows you to control whether the incoming information is going to be accumulated or not. There is also an oblivion door, which allows us to determine if the state of the previous cell will be taken into account. In addition, an output cell determines whether the output of the cell is to be propagated to the final state. The cell type and control gates are the most important advantage of the LSTMs since they get the gradient trapped in the cell and prevent it from disappearing too fast, which is a critical problem of the basic RNN model [17]. Xingjian et al. [15] developed a multivariable version of LSTM that they called ConvLSTM, which has convolutional structures in both state-to-state and state-to-state transitions. By stacking multiple ConvLSTM layers, they were able to build a model that proved to have excellent spatio-temporal sequence prediction capability, suitable for making predictions in complex dynamic systems.

Next Route Routing (NRR) [18] extends the functionality of the SCATS system (Sidney Coordinate Adaptive Traffic Systems) by adding vehicle congestion and re-routing detection mechanisms. The re-routing decision takes into account the destination of the trip and the current local traffic conditions. The proposed architecture allows the positive impacts of re-routing to extend over a larger area. The results obtained show a reduction of 19% in the ATT. It is necessary to highlight two points on this proposal. First, it is an extension of a previous system (SCATS), it is not an autonomous system. Second, it requires all vehicles to provide information to know the actual traffic status.

3   Proposed Solution

The innovation of this proposal is in the combination of three elements. First, the system does not require that all vehicles constantly send information to detect congested areas due to the combined use of historical data and the predictions made by the model of Deep Learning. The model uses a deep ConvLSTM architecture that has been shown to able to handle problems prediction spatio-temporal sequences similar to the prediction of vehicular traffic [15]. Secondly, the use of the Entropy-Balanced k-Shortest Path algorithm to generate alternative routes, which allows creating a global strategy to manage congestion. Third, and finally, the information obtained from the probe cars to detect early non-recurring congestion.

3.1   Architecture Proposed

The proposed architecture takes into account that traffic is a dynamic problem and with a geographical distribution in which several autonomous entities interact. Considering this, it was chosen an architecture based on agents. The use of agents significantly improves the analysis and design of problems similar to the problem of traffic detection [19]. In addition, the need for dynamic traffic management involving a large number of vehicles requires more reactive agents than cognitive agents. Unlike cognitive agents, reactive agents have no representation of their environment. Reactive agents can cooperate and communicate through interactions (direct communication) or through knowledge of the environment.

Vehicle Agent (VA) is responsible for requesting the best possible route to the Route Guidance Agent (RGA). One in 10 vehicles is designated as the Probe Car Agent (PCA). These agents have an additional task, the responsibility of reporting at each intersection the delay experienced when traveling that street to the corresponding Cruise Agent (see Fig. 1). The Cruise Agent (CA) collects data from the PCAs that cross the streets it controls. Every 5 min [20], it uses the collected data to generate a report, which is sent to the Zone and Network Agents.

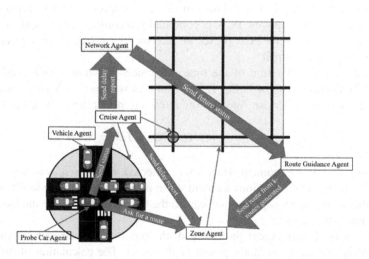

Fig. 1. As soon a Probe Car Agent crosses an intersection, it sends to the Cruise Agent the status of the street. Every 5 min, different reports are exchanged between Zone, Network and Route Guidance Agents.

The Zone Agent (ZA) detects congestion in an area of the city and sends alert to vehicles. The Network Agent (NA) uses a combined scheme of historical data and a Deep Learning model to predict the future state of the network during the next 5 min. Lastly, Route Guidance Agent (RGA) generates alternative routes.

Beginning a trip

At the beginning of the trip, the Vehicle Agent sends a Route-Suggestion message to the Zone Agent. This message contains its id, the origin, and the destination of its trip. The Zone Agent forwards this message to the Route Guidance Agent. The RGA verifies if there are alternatives generated. If there is none, it generates a set of k-Shortest Paths using a multithreaded algorithm [21]. With the set of k-Shortest Paths, the Route Guidance Agent sends the shortest route less suggested until that moment to the Zone Agent. The Zone Agent forwards the response to the VA, but first the route the vehicle will follow. This information will be used to send alert messages when congestion is detected in the area.

A similar exchange of messages is made between the Probe Car, Zone, and Route Guidance Agents; except that when Zone Agent forwards the route, it adds a list of Cruise Agents to which the Probe Car Agent must report the situation of the streets through which it crosses.

Predicting the State of the Vehicular Network

Probe Car Agents must report to Cruise Agents the delay experienced when traveling one of the streets that CA controls. The algorithm used to calculate the delay is based on the algorithm defined by Chuang [22] and is modified to detect and report non-recurrent bottlenecks. The idea behind this algorithm is that a trip is an alternate sequence of periods in which the car is stationary and periods in which it is moving. A STOP event indicates when a vehicle has moved from a complete stop and can be described with the GPS position and the time the event occurred. A GO event indicates when a stopped vehicle moves. PCA is continually recording events until it reach an intersection. As soon as the PCA passes an intersection, it calculates the delay it got when traveling on that street.

The total delay is the sum of the time differences between a GO event and the previous STOP event as show in Eq. 1. The result, as well as the identifier of the street segment, are sent to the Cruise Agent of the intersection that the PCA has just crossed.

$$Delay_{vehicle} = \sum_{j=2} \left(TIME\_GO_j - TIME\_STOP_{j-1} \right) \tag{1}$$

where $TIME\_GO_j$ is the moment when it was detected that *vehicle* was moving again, and $TIME\_STOP_{j-1}$ is the previous moment when *vehicle* stopped. The result, as well as the identifier of the street segment, are sent to the Cruise Agent of the intersection that the PCA has just crossed.

Every 5 min, Cruise Agent processes all the reports received and determines the average delay that exists in that segment of the street. The calculation of the average delay is based on Eq. 2.

$$Delay_{avg} = \frac{\sum_{i=1}^{N} Delay_{vehicle_i}}{N} \qquad (2)$$

where i refers to any probe car that has passed through that particular intersection during the last 5 min interval, N is the total probe cars that have passed and $Delay_{vehicle_i}$ is the delay reported by $Vehicle_i$ at the time of passing through that intersection, calculated using Eq. 1. The Cruise Agent sends an Update message to the Network Agent. This message contains a timestamp, ids of the streets it controls, as well as the average delay of each of the streets.

The Network Agent uses these messages to generate the next forecast (see Algorithm 2). The process is as follows: if there is no Update message for a given street, the proposal uses Eq. 3 to generate an expected delay:

$$Delay_{street} = \alpha \times Historical\_Value_{street} + (1 - \alpha) \times Prediction\_Value_{street} \qquad (3)$$

where α indicates the weight assigned to the historical data, $Historical\_Value$ is the historical data reported for that *street* segment during the last 5 min interval and *Prediction_Value* is the value predicted by the model for that *street* segment during the last 5 min interval. This will allow us, later, to change the value and observe what is the proportion of historical data and forecast that generates a greater reduction of the Average Travel Time (ATT). Once the process is finished, the Deep Learning model uses the data obtained to generate a new forecast. This forecast will be valid for the next 5 min. If a Non-Recurrent-Congestion message is received during this time, the message will be stored for later use if a more recent Update message does not arrive. Lastly, Network Agent sends the new forecast to both Zone and the Route Guidance Agents.

The Zone Agent analyzes the forecast in search of streets that may have an average delay greater than the limit determined by the system. Once these streets are detected, the Zone Agent sends a Recurrent-Congestion message to vehicles that have one or more of these streets on their route. The Zone Agent sends a Route-Update message to the Route Guidance Agent and all Vehicle Agents who have one or more of the congested streets on their routes. Any VA (or PCA) that receives a Route-Update message should verify how far it is from the congested highway. If the distance is less than a limit defined by the system, it will send a Route-Suggestion message to the Zone Agent. In this case, the message will indicate the current position of the vehicle as point of origin. Meanwhile, as soon as RGA receives the Route-Update message, it invalidates all previously generated k-Shortest Paths to generate new ones.

With regard to non-recurrent congestion, the Probe Car Agent carries out detection. If the PCA detects an excessive number of STOP events while crossing a street, it will proceed to send a Non-Recurrent-Congestion message to both the Zone Agent and the Network Agent. This message contains a timestamp, street id and delay experienced so far. The Zone Agent forwards a Non-Recurrent-Congestion message to the Route Guidance Agent and to all Vehicle Agents that that street has on its route. In the case of the RGA, it will verify if the message has updated information and, if so, it will invalidate only the k-Shortest Path containing the id of the congested street and update the information of that street in its forecast. Any VA (or PCA) that receives a Non-

Recurrent-Congestion message should verify how far away it is from the congested road. If the distance is less than a limit defined by the system, it will send a Route-Suggestion message to the Zone Agent. In this case, the message will indicate the current position of the vehicle as point of origin. For its part, the NA will keep this report for later use, during the process of generating a new forecast.

4   Performance Evaluation

4.1   Tools Used

For traffic simulation, we mainly use two tools: SUMO and TraCI4J. SUMO [23] is an open source program for urban traffic simulation. It is a microscopic simulator; that is, each vehicle is modeled explicitly, has its own route and moves independently through the network. It is mainly developed by the Institute of Transportation Systems. The version used in this work was version 0.25.0, published in December 2015. TraCI4J [24] is an API developed for SUMO that allows communication in time of execution between the proposed system and the simulator. TraCI4J was developed by members of ApPeAL (Applied Pervasive Architectures Lab) of Politecnico di Torino. This API offers a higher level of abstraction and better interoperability than other APIs developed for the same task.

Additionally, we use Keras for the implementation and training of Deep Architecture. Keras [25] is a high-level neural network API, developed in Python and capable of running on TensorFlow, CNTK or Theano. In our case, and with the intention of taking advantage of a server with two NVidia GTX 1080 cards, we chose to use TensorFlow [26] as a backend.

4.2   Test Scenario

After reviewing previous works, we observed that one of the most used scenarios to evaluate proposals is the scenario known as Manhattan (or grid type) [1, 27]. In our case, two synthetic networks were designed based on the Manhattan scheme: one of the 4×4 (M4) and another 5×5 (M5). For example, 5×5 means that this map has five intersections on the vertical and horizontal axes. All segments of the street have a length equal to 200 m and each of them consists of two lanes with opposite directions.

4.3   Simulation Setup

SUMO includes a set of applications that help prepare and perform the simulation of a traffic scenario. A simulation scenario can contain a large number of vehicles. Even for small areas, such as the test traffic networks, it is almost impossible to define traffic demand manually. For this reason, we use "randomTrips.py" [28] and duarouter [29], which are included in the SUMO toolkit. The "randomTrips.py" tool generates a set of random trips choosing an origin and a destination evenly at random. The resulting trips are used by duarouter to define the route that each vehicle will use to reach its destination. The algorithm used to generate this route is the Dijkstra [30] algorithm. Because of this process, each vehicle is provided with a unique identifier, departure time and vehicle route through the road network. SUMO uses this file during the simulation.

The file defines a full day of traffic with three traffic flows: LIGHT (from 5% to 20% of maximum capacity), MEDIUM (from 60% to 70% of maximum capacity) and HEAVY (from 80% to 95% of maximum capacity). The distribution of traffic flows is similar to that shown in Fig. 2. Each intersection of the synthetic scenarios has semaphores that use a static configuration, that is, each semaphore maintains the same duration in each phase, independently of the changes in traffic conditions.

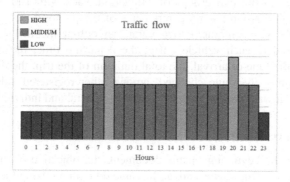

Fig. 2. Traffic flows simulated by time of day

We perform 2,400 simulations in order to obtain the data to train the deep network architecture. Since the results of these simulations will be used for deep network training, no component of the proposal was used. During these simulations, all the vehicles reported the delay in experiencing the streets of their route (see Eq. 1). With these reports, every 5 min, the average delay experienced in each of the streets (see Eq. 2) was calculated, generating a delay matrix with these values. At the end of a full simulation day, a file containing 288 matrices of this type was generated. This file, therefore, contains the behavior of the traffic network throughout the 24 h of the simulation (1,440 min/5 min). Finally, all values were normalized using the Min-Max technique (see Eq. 4).

$$Z_i = \frac{x_i - \min(x)}{\max(x) - \min(x)} \tag{4}$$

With the files obtained from the simulations, a dataset was constructed, which was divided into two groups: training (80%) and validation (20%).

The initial architecture had two internal ConvLSTM layers. This architecture was implemented and evaluated, obtaining 75% accuracy. To increase the accuracy of the model, a new intermediate layer ConvLSTM was added. This process was repeated twice more until 88% accuracy was obtained. The resulting network has four internal ConvLSTM layers (see Fig. 3). This architecture proved to have a more accurate spatio-temporal sequence prediction capability, which makes it suitable to provide predictions in complex dynamic systems such as the problem of traffic flow prediction.

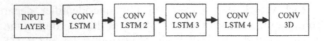

Fig. 3. Final deep architecture

From the initial group of 2,400 simulations, we randomly selected 1,000. The results of these simulations were used to generate the historical information that the proposal requires. In order to obtain baselines to determine the efficiency of the proposal, we decided to run two groups of simulation, each with 100 simulations were carried out. The first group has no components of the proposal were used. However, this time, SUMO was configured to generate an output file that contained all the information related to each vehicle's trip: the departure time, the time the vehicle wanted to start, the time of arrival, the total duration of the trip, the distance traveled, the number of times the route changed, among other registered values. With these values as a basis, it was possible to determine both general and individual performance.

4.4   Performance Metrics

As mentioned at the beginning of this document, the objective of the proposal is to alleviate congestion in the streets and, at the same time, minimizing the Average Travel Time (ATT). The ATT is calculated using Eq. 5. We use this metric because its reduction leads to lower fuel consumption, higher economic growth and better living experience for citizens [31].

$$Time_{avg} = \frac{Time_{sum}}{K} = \frac{\sum_{i=1}^{K} Time_{vehicle_i}}{K} \qquad (5)$$

where i refers to each of the vehicles that were generated for the simulation, $Time_{Vehicle_i}$ is the total travel time of $Vehicle_i$ and K is the total vehicles in the simulation.

5   Results

The second group were executed, this time using the components of the proposal with the parameters presented in Table 1. It is also necessary to highlight that:

1. 10% of the vehicular population will be Probe Car Agents.
2. All the simulations continue until all vehicles complete their trips.
3. The value of $\alpha = 0.5$, which implies using a mix 50-50 of historical data and predictive model.

Table 1. Parameters used in evaluation

Parameter	Values
Period	Frequency of re-routing; by default period = 5 min
Threshold	Maximun delay per street; by default = 25 s
Level L	Distance that should be the vehicles of the congested segment to request re-routing; by default $L = 5$
α	Weight that will be given to historical data; by default $\alpha = 0.5$
k-Paths	Maximum number of alternative paths for each vehicle; by default $k = 5$

In the case of the M4 network, as we can see in Fig. 4a that there are significant reductions. The average duration of travel time without using the proposal was 218.49 s; while using the proposal, the average was 203.70 s was obtained. Which implies a 6.77% reduction. Similarly, in terms of waiting time (the time in which the vehicle speed is less than 0.1 m/s) went from 94.85 to 82.59 s when the proposal is used, which implies a reduction of 12.95%. Finally, we can also see a reduction of 10.15% in terms of time loss (the time required due to driving below the ideal speed), going from 147.21 to 132.27 s.

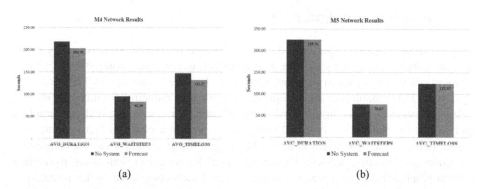

Fig. 4. Result obtained for (a) M4 network, without using the proposal (No System) and using it (Forecast); (b) M5 network, without using the proposal (No System) and using it (Forecast)

With respect to the M5 network, although reductions were achieved in the three aforementioned areas, they were not as favorable as in the M4 network (see Fig. 4b). Regarding the average duration, it went from 225.82 to 225.76 s, which means a reduction of only 0.03%. The waiting time went from 75.83 to 75.67 s, which is a reduction of 0.21%. Finally, the time loss got a reduction of 0.20%, from 123.61 to 123.37 s.

Figure 5 show the histograms of the improvements of ATT under normal traffic conditions when we use the proposed system. In this case, the improvement ratio is defined as *ratio = oldtime/newtime*. For example, a ratio of two means that the vehicle was able to complete its trip in half of the time, while a ratio of three means that it finished its trip in one third of the previous time. As we observed, a large number of vehicles, 50% (M4) and 56% (M5), saved 20% of the travel time (ratio 1.25). Although we can also observe that, there are vehicles that could avoid the worst part of the traffic by completing their trip two or three times faster (relations 2 and 3).

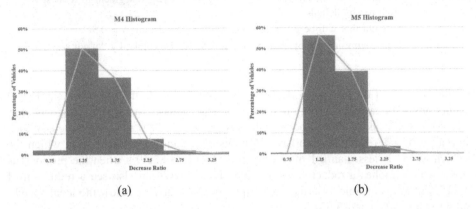

(a)   (b)

Fig. 5. Histogram of gain in travel times obtained for (a) M4 network and (b) M5 network

6 Conclusions and Future Work

The congestion caused by public and private transport is the main cause of many of the problems that cities face today. The best approach to solve this congestion is a more efficient use of the installed road infrastructure. The use of Intelligent Transportation Systems (ITS) technologies allows optimizing the route, reducing distances, increasing safety in transport, reducing the time spent in traffic jams, reducing fuel consumption and improving air quality. In this document, we present a platform that continuously monitors the traffic flow status in a specific geographic area and provides a congestion detection and warning service. This allows vehicles to adjust their route from their current point to their destination when the system detects congestion on this route, even when the vehicle is already moving. The proposed system also considers those situations for which it is not possible to obtain updated information, using a real-time prediction model based on a deep neural network. The results obtained from a series of simulations carried out on two synthetic scenarios show that the proposal is capable of generating improvements in ATT for each of the scenarios analyzed, achieving better results than other existing proposals.

Seeing how different were the results obtained for the two test networks, the possibility of modifying the algorithm to generate a new forecast was raised. The change is to exclude historical data; we will only use the data in real time and the previous forecast. The simulations are currently being generated using the modified algorithm. Also, it has also been considered to evaluate the proposal with real data. For this, we have already selected a sub-scenario of TAPASCologne 0.17.0 [32]. TAPAS Cologne is an open project that provides a large scale data set that allows you to make, using SUMO, a very realistic urban vehicle simulation. This scenario uses a real

Colonia map drawn from OpenStreetMap [33] and generates traffic control from 6:00 AM at 8:00 AM, Using the methodology of simulation of travel and activity patterns (TAPAS) and dynamic allocation algorithm called Gawron.

References

1. Wang, S., Djahel, S., McManis, J.: A multi-agent based vehicles re-routing system for unexpected traffic congestion avoidance. In: 2014 IEEE 17th International Conference on Intelligent Transportation Systems (ITSC), pp. 2541–2548. IEEE (2014)
2. de Souza, A.M., Yokoyama, R.S., Maia, G., Loureiro, A., Villas, L.: Real-time path planning to prevent traffic jam through an intelligent transportation system. In: 2016 IEEE Symposium on Computers and Communication (ISCC), pp. 726–731. IEEE (2016)
3. Waze. https://www.waze.com/es-419. Accessed 01 Apr 2006
4. Google Maps. https://www.google.com.mx/maps. Accessed 01 Apr 2019
5. Pan, J., Khan, M.A., Popa, I.S., Zeitouni, K., Borcea, C.: Proactive vehicle re-routing strategies for congestion avoidance. In: 2012 IEEE 8th International Conference on Distributed Computing in Sensor Systems, pp. 265–272. IEEE (2012)
6. Song, Q., Wang, X.: Efficient routing on large road networks using hierarchical communities. IEEE Trans. Intell. Transp. Syst. 12(1), 132–140 (2011)
7. Schmitt, E., Jula, H.: Vehicle route guidance systems: Classification and comparison. In: Intelligent Transportation Systems Conference, 2006, ITSC 2006, pp. 242–247. IEEE (2006)
8. Khanjary, M., Faez, K., Meybodi, M.R., Sabaei, M.: Persiangulf: an autonomous combined traffic signal controller and route guidance system. In: 2011 IEEE Vehicular Technology Conference (VTC Fall), pp. 1–6. IEEE (2011)
9. Bauza, R., Gozálvez, J.: Traffic congestion detection in large-scale scenarios using vehicle-to-vehicle communications. J. Netw. Comput. Appl. 36(5), 1295–1307 (2013)
10. Araújo, G.B., Queiroz, M.M., de LP Duarte-Figueiredo, F., Tostes, A.I., Loureiro, A.A.: Cartim: a proposal toward identification and minimization of vehicular traffic congestion for vanet. In: 2014 IEEE symposium on computers and communications (ISCC), pp. 1–6. IEEE (2014)
11. Meneguette, R.I., Ueyama, J., Krishnamachari, B., Bittencourt, L., et al.: Enhancing intelligence in inter-vehicle communications to detect and reduce congestion in urban centers. In: 20th IEEE symposium on computers and communication (ISCC), Larnaca, 6–9, pp. 662–667 (2015)
12. de Souza, A.M., Yokoyama, R.S., Botega, L.C., Meneguette, R.I., Villas, L.A.: Scorpion: a solution using cooperative rerouting to prevent congestion and improve traffic condition. In: 2015 IEEE International Conference on Computer and Information Technology; Ubiquitous Computing and Communications; Dependable, Autonomic and Secure Computing; Pervasive Intelligence and Computing, pp. 497–503. IEEE (2015)
13. Kumar, S.V.: Traffic flow prediction using kalman filtering technique. Procedia Eng. 187, 582–587 (2017)
14. Ma, X., Dai, Z., He, Z., Ma, J., Wang, Y., Wang, Y.: Learning traffic as images: a deep convolutional neural network for large-scale transportation network speed prediction. Sensors 17(4), 818 (2017)
15. Xingjian, S., Chen, Z., Wang, H., Yeung, D.Y., Wong, W.K., Woo, W.C.: Convolutional LSTM network: a machine learning approach for precipitation nowcasting. In: Advances in Neural Information Processing Systems, pp. 802–810 (2015)

16. Hochreiter, S., Schmidhuber, J.: Long short-term memory. Neural Comput. **9**(8), 1735–1780 (1997)
17. Pascanu, R., Mikolov, T., Bengio, Y.: On the difficulty of training recurrent neural networks. In: International Conference on Machine Learning, pp. 1310–1318 (2013)
18. Wang, S., Djahel, S., Zhang, Z., McManis, J.: Next road rerouting: a multiagent system for mitigating unexpected urban traffic congestion. IEEE Trans. Intell. Transp. Syst. **17**(10), 2888–2899 (2016)
19. Adler, J.L., Blue, V.J.: A cooperative multi-agent transportation management and route guidance system. Transp. Res. Part C Emerg. Technol. **10**(5), 433–454 (2002)
20. Brennand, C.A., de Souza, A.M., Maia, G., Boukerche, A., Ramos, H., Loureiro, A.A., Villas, L.A.: An intelligent transportation system for detection and control of congested roads in urban centers. In: 2015 IEEE Symposium on Computers and Communication (ISCC), pp. 663–668. IEEE (2015)
21. Murueta, P.O.P., Pérez, C.R.C., Uresti, J.A.R.: A parallelized algorithm for a real-time traffic recommendations architecture based in multi-agents. In: 2014 13th Mexican International Conference on Artificial Intelligence (MICAI), pp. 211–214. IEEE (2014)
22. Chuang, Y.T., Yi, C.W., Lu, Y.C., Tsai, P.C.: iTraffic: a smartphone-based traffic information system. In: 2013 42nd International Conference on Parallel Processing (ICPP), pp. 917–922. IEEE (2013)
23. SUMO – Simulation of Urban Mobility. http://sumo.sourceforge.net/. Accessed 03 Sept 2017
24. Gueli, E.: TraCI4J. https://github.com/egueli/TraCI4J. Accessed 03 Sept 2017
25. Keras: The Python Deep Learning library. https://keras.io/. Accessed 03 Sept 2017
26. Tensorflow. https://www.tensorflow.org/. Accessed 03 Sept 2017
27. de Souza, A.M., Yokoyama, R.S., Boukerche, A., Maia, G., Cerqueira, E., Loureiro, A.A., Villas, L.A.: Icarus: Improvement of traffic condition through an alerting and re-routing system. Comput. Netw. **110**, 118–132 (2016)
28. randomTrips. http://sumo.dlr.de/wiki/Tools/Trip. Accessed 03 Sept 2017
29. DUAROUTER. http://sumo.dlr.de/wiki/DUAROUTER/. Accessed 03 Sept 2017
30. Dijkstra, E.W.: A note on two problems in connexion with graphs. Numer. Math. **1**(1), 269–271 (1959)
31. Rao, A.M., Rao, K.R.: Measuring urban traffic congestion-a review. Int. J. Traffic Transp. Eng. **2**(4) (2012)
32. Data/Scenarios/TAPASCologne. http://sumo.dlr.de/wiki/Data/Scenarios/TAPASCologne/. Accessed 03 Sept 2017
33. Haklay, M., Weber, P.: Openstreetmap: User-generated street maps. IEEE Pervasive Comput. **7**(4), 12–18 (2008)

From a Conceptual to a Computational Model of Cognitive Emotional Process for Engineering Students

Sonia Osorio Angel, Adriana Peña Pérez Negrón[✉],
and Aurora Espinoza Valdez

Computer Science Department, CUCEI, Universidad de Guadalajara,
Blvd. Marcelino García Barragán #1421, 44430 Guadalajara, Jalisco, Mexico
{sonia.oangel, aurora.espinoza}@academicos.udg.mx,
adriana.pena@cucei.udg.mx

Abstract. Recently, research in Education has focused on analyzing emotions as an important factor that influences the processes related to learning, beyond academic knowledge and domain contents. Education studies are now considering emotions in students experience during their learning process, their positive or negative influence for better academic support. In this context, Affective Computing aim is to create computer environments for learning capable of promoting positive emotional experiences that improve student's performance. We are proposing a conceptual model for the analysis of emotions in Spanish language, intended for Engineering students during their learning process in a collaborative computer environment. The model was designed through carrying out an exploratory study with an interpretative descriptive analysis in order to find patterns in the empirical evidence that in the future leads to the automation of the phenomenon using Affective Computing resources.

Keywords: Affective Computing · Sentiment Analysis · Cognitive process

1 Introduction

Education has largely evolved to the accelerated advance of technology. The old traditional educational model in which the protagonist teacher dictated a lecture, while the students listened and repeated is becoming obsolete. Today, education scenarios should be built under a social perspective. In this context, nowadays students are part of a generation for which information flows through various means and devices available on a daily basis. Therefore, we must take advantage teaching through technological advances that promote the social aspect.

Also, one of the main arguments in new educational proposals or models is taking into account emotions and their direct influence on the cognitive process. Emotional Intelligence (EI) is a field recently explored in which, according to Goleman [1], the student requires skills such as self-control, enthusiasm, empathy, perseverance and the ability to motivate oneself, which represents to be aware of our emotions and understand the others feelings. These skills can be configured genetically, be molded on the

© Springer Nature Switzerland AG 2020
J. Mejia et al. (Eds.): CIMPS 2019, AISC 1071, pp. 173–186, 2020.
https://doi.org/10.1007/978-3-030-33547-2_14

first years, or can be learned and perfected throughout life. Furthermore, Goleman [1] argues that EI is more important than IQ. Students who are anxious, angry, or depressed do not learn; people caught in these emotional states do not get information efficiently or do not deal with it in a proper way. In other words people with emotional intelligence are more likely to achieve excellent school performance and its subsequently success at work.

In addition to the pedagogical and psychological studies exist serious efforts to create integral and holistic computational models, capable of combining emotional and academic education. This is related to a relatively recent research area called Affective Computing (AC), which objective is to explore the human affective technology experience through the combination of several disciplines such as Engineering, Psychology, Cognitive Science, Neuroscience, Sociology, Education, Psychophysiology, and even Ethics [2].

According to Kort [3] AC for Education faces two main problems, investigating a new educational pedagogy, and the construction of computerized mechanisms that recognize accurately and immediately the student status by some ubiquitous method, to activate a proper response. We believe that these new pedagogies should take into account the cognitive styles that students have, since there are different theories indicating that according to the area of study, students have different cognitive characteristics. In [4] it is claimed that in Science, Mathematics, Engineering, and Technology (SMET) Education there is a great need to provide students with more than just technical skills, in these particular areas the students are often known for their ability to solve problems and their inability to work with other people, which is likely to affect their emotional state, that is, the emotions an engineering student experiences are not exactly the same as those felt by, for example, a social science student. Our work focuses on engineering students, a model that responds to their particular characteristics and needs.

Within the aim of AC, our proposal is based on a conceptual model emerged from an exploratory study carried out with engineering students through a virtual environment in discussion forums. According to Tromp [5] the design of an application to support collaboration requires the understanding of real world collaboration in such a way that the computer tools can be explicit. In our case, from real world situations can be inferred indicators to evaluate emotions that might affect the cognitive process, and to understand whether or not effective learning is achieved. The conceptual model consists on analyzing the dialogues in Spanish language generated by students in a learning context, looking for evidence of cognitive and emotional aspects manifested in text. This model is then raised from a computational perspective through Affective Computing. However, AC involves multiple methods and techniques for the study of emotions such as body movements, facial expressions, and text analysis among others [6]. Within AC the subarea Sentiment Analysis (SA) or Opinion Mining (OM) has as main target the computational treatment of opinions, feelings and subjectivity towards an entity; all of them expressed by computer users through their written text on the multiple platforms of the Web. Therefore, SA was considered as the classification process.

This paper is organized as follows: Sect. 2 describes related work to cognitive and emotional models, as well as Sentiment Analysis systems for Spanish in the Education area. Section 3 the background, cognitive and emotional models that support our proposal. Section 4 details the exploratory study model and its results. In Sect. 5 an approach on how to automate the model is presented, and finally discussion and future work is in Sect. 6.

2 Related Work

There are several models focused on analyzing the quality of students' interactions in virtual environments in order to assess their learning level manifested through text. These works focus is the Content Analysis of transcriptions, establishing coding schemes through discursive categories and indicators, which provide researchers with a framework to observe how knowledge is constructed through discourse. Of course, the learning process is complex and there are many factors that must be taken into account, that is why there are different proposals which in some cases do not match on the indicators to perform Content Analysis [7]. Among them, Henri [8] argued that learning is accomplished through 5 dimensions or categories: participative, social, interactive, cognitive and metacognitive. His model combines qualitative and quantitative strategies measuring mainly participant's interaction.

In [6] it was pointed out that the number of interactions does not provide enough information to evaluate if there is social construction of knowledge. Therefore, they propose parameters to validate this construction. These parameters correspond to 5 phases or categories: I. Sharing/comparing Information, II. Discovering/exploring dissonance, III. Knowledge negotiation, IV. Testing/modifying proposal construction, and V. New knowledge/understanding/perspectives.

Another proposal was presented in [9], where the authors, encouraging empirical findings, related to an attempt to create an efficient and reliable instrument to assess the nature and quality of discourse and critical thinking in a text-based educational context. In their work established that the education experience as a system is constituted by three dimensions: (a) Cognitive presence, (b) Social presence, and (c) Teaching presence. From these dimensions or categories, they developed a series of indicators to observe cognitive presence in the textual student's interactions.

In [10] a system was developed with a large number of categories which are designed through three dimensions: social, cognitive and didactic. This proposal is based on [9], but adapted to their study requirements. Finally, Rourke [11] presented main problems related to content analysis, documenting their evolution through a review of 19 computer mediated communication content analysis studies, in which they included their own work [12]. The paper presents criteria to validate if descriptions or relationships resulting from content analysis are valid, how content analysis has been characterized, and the types of content and the analysis units used. Additionally, it explores ethical issues and software available to support content analysis.

Also, there are several emotional theories that provide a solid foundation for working computationally emotional models. Among others, the most used are the taxonomic model of Plutchik [13], which establishes 8 primary emotions in 4 axes: joy-sadness, disgust-acceptance, anger-fear, and surprise-anticipation. Plutchik makes combinations of two primary emotions giving rise to secondary emotions, for example, acceptance + fear = submission or joy + acceptance = love.

In [14] Russell presented a Circumplex model of affect and the theory of pleasure-activation, where emotions are described in two dimensions: pleasure-displeasure and activation-deactivation.

Ekman [15], stated that there are six basic emotions: joy, sadness, fear, disgust, surprise, and anger, that through facial expressions are manifested and recognized in all cultures to similar stimuli.

In [3] Kort presents a model with emotions experienced by students in technology area. The model has six axes of emotion, which can arise in the course of learning and that go from negative to positive states.

Regarding the SA on Education and specifically for Spanish language, in [16] is presented a Facebook™ application that retrieves users' written messages and classifies them according to their polarity (i.e. positive, negative and neutral), showing the results through an interactive interface. The application detects emotional changes in periods of time through statistics, and it also seeks emotions from friends. They used a hybrid classification method by combining automatic and lexical-based learning techniques.

In [17] a novel approach is proposed to label the affective behavior in educational discourse based on fuzzy logic that is lexicon based, which allows a human or virtual tutor to capture the student's emotions, making the students aware of their own emotions, evaluating these emotions and providing affective appropriate feedback. Their proposal is a diffuse classifier that provides a qualitative and quantitative evaluation. Their approach has been proven in a true online learning environment and they claim that it had a positive influence on the students' learning performance.

In [18], an intelligent affective tutor based on SA is presented. This system takes students' opinions regarding programming exercises, and then performs the analysis by determining if the polarity of the exercise is positive (like) or negative (not like). They used Naïve Bayes classifier, arguing that it is the simplest and most used classifier.

Our proposal is based on the Content Analysis method proposed by Garrison, Anderson and Archer [9]. The content of communication or our interest is both cognitive presence and emotions, which will be automated using SA techniques. We believe that our proposal goes beyond observing whether the student feels good or not, or if the exercise was or not liked. Next, the theoretical models that support our proposal are described.

3   Theoretical Models that Support the Proposal

This proposal is build up on two models. The Community of Inquiry Model proposed by Garrison, Anderson and Archer [9] that provides the theoretical framework and tools for Computer Mediated Communication (CMC), suitable for education in a virtual environments based on text. They offer a template or tool for researchers to analyze written transcripts, as well as a guide to educators for the optimal use of computer conferencing as a medium to facilitate an educational transaction. And, regarding the emotional factor, the model presented in [19] by Kort for emotions experienced especially in students in Science, Mathematics, Engineering and Technologies areas, which are the areas where our teaching activity develops.

3.1 The Community of Inquiry Model

This model provides a useful methodology for Content Analysis for textual messages, with categories and indicators that make possible to evaluate whether learning is taking place in the CMC environment. The Community Inquiry Model establishes that three elements must be manifested to guarantee an educational experience: Cognitive Presence, Social Presence and Teaching Presence. In Fig. 1 are presented these elements with its categories, which are the basis for messages classification.

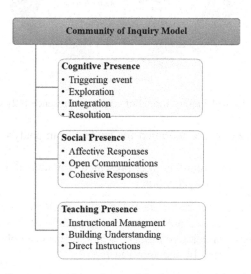

Fig. 1. Community of inquiry model: elements and its categories.

For the current study we considered only Cognitive Presence for Content Analysis. We assume that Social Presence has been manifested because forum activities were designed to be solved through teamwork. Also, the Teaching Presence was obviated, because professor participated in the activity design, and he/she mediated the discussion forums.

The Community of Inquiry Model, presents a submodel denominated Practical Inquiry Model to assess and guide dialogic writing for the purpose of creating Cognitive Presence [12]. This model focuses on higher-order thinking processes as opposed to specific individual learning outcomes, and it is a generalized model of critical thinking. The critical thinking perspective employed is comprehensive and it includes creativity, problem solving, intuition, and insight. To make operational cognitive presence, the Practical Inquiry model has been divided in four phases (quadrants), see Fig. 2, that correspond to stages which describe the Cognitive Presence in an educational context:

Quadrant 1. *Triggering event*, the phase in which a problem, issue or dilemma is identified.
Quadrant 2. *Exploration*, divergent phase characterized by brainstorming, questioning and information exchange.

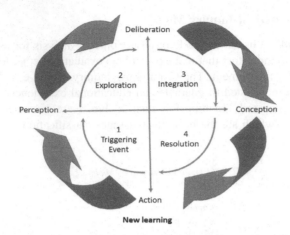

Fig. 2. Practical inquiry model of cognitive presence [12], adapted.

Quadrant 3. *Integration*, phase in which the students can analyse and synthesize relevant information.

Quadrant 4. *Resolution*, solutions are presented implementing solutions or test hypothesis through practical application.

The phases of the model are the ideal logical sequence of the process for critical inquiry and, therefore, they are not immutable. In addition to the original model, we consider that learning is an iterative process, which restarts the phases with every new learning episode, represented by the spiral in Fig. 2.

3.2 Emotions and Learning Interaction Model

To analyse the students' affective state it was selected the model proposed in [19] and [3]. It consists on a set of emotions with six axes, which arise in the learning process in students of the SMET areas. The six axes go from the negative (left) to the positive (right): anxiety to confident, tedium to fascination, frustration to euphoria, discouragement to enthusiasm, terror to excitement, humiliation to pride, passing through a number of emotional states. Figure 3 shows the six emotions axes in the way they were distributed in each quadrant in the Kort model (see Fig. 4).

1	Ansiety	Worry	discomfort	Confort	Hopeful	Confident
2	Ennui	Boredom	Indifference	Interest	Curiosity	Intrigue
3	Frustration	Puzzlement	Confusion	Insight	Enlightenment	Epiphany
4	Dispirited	Dissapointed	Disatisfied	Satisfied	Thrilled	Enthusiastic
5	Terror	Dread	Apprehension	Calm	Anticipatory	Excited
6	Humilliated	Embarrassed	Self-conscious	Pleased	Satisfied	Proud

Fig. 3. Set of emotions in learning process.

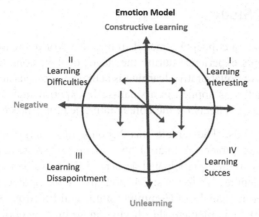

Fig. 4. Model of interaction between emotions and learning (adapted form [19]).

The flow among quadrants in the Kort model goes in opposite direction to the model of Practical inquiry of the cognitive presence. However, in a exploratory study described in the next section, it was observed that in order to achieve learning is necessary for the student to travel through the four quadrants of the Practical inquiry model, while in the emotion model it is not.

In fact, the authors clarify that at any given time, the student may be in several quadrants of the Kort model, with respect to different axes. The students might be, for example, in quadrant II feeling frustrated and simultaneously in quadrant I with regarding interest. It is also important to recognize that a range of emotions occurs naturally in a real learning process, and it is not always the case that positive emotions are good. The model does not try to keep the student in quadrant I or IV, but rather to help him see that the cyclical nature of learning in the SMET areas implies landing on the negative half, as an inevitable part of the cycle. Figure 4 shows the Kort model quadrants with an assigned name for the exploratory study. Also, internal arrows were added that represent the states in which the student might transit from one quadrant to another.

In each quadrant a series of emotions correspond to certain stages during the learning process. In quadrant I, there are emotions related to involvement, exploratory learning and discovery, that is, interest and curiosity to learn, it was decided to call it *Learning interesting*. In *Learning Difficulties* quadrant II, the negative emotions corresponding to the difficulties derived from misconceptions or incomplete comprehension; here the Teaching Presence must intervene so that the negative emotions do not persist. In quadrant III, *Learning Disappointment*, the student recognizes working based on an erroneous or incomplete idea. In this stage the intervention or Teaching presence must provide the necessary emotional support to get out of the disappointment, or any other negative emotion to regain the learning cycle. The authors stress that it must not be forgotten that some students require more emotional support than others, and that the Teaching presence in quadrant III is possibly the most challenging one. In quadrant IV, *Learning Success*, the student regains hope and positive attitude, he/she can feel proud and satisfied of achieving learning, and then the cycle starts again.

4   Exploratory Study

With the aim to create a conceptual model from a real world situation, for identifying emotions that students experience during the stages of their cognitive process, and to evaluate if an effective collaborative learning is achieved, a discussion forums was used as tool. The approach was applied in a small-scale experimental learning situation to analyze and evaluate its effects on the collaborative learning process.

Methods and Materials. The sample was 2 groups of undergraduate students in the Computer Programing subject. A total of 56 students, 10 women and 46 men, in the range of 18 to 24 years old, participated as part of their evaluation. The students are studying different degrees as follows: 34 students of Computer engineering, 11 of Informatics engineering, and 11 of Communications and Electronics engineering. Two platforms were used to implement the discussion forums. Moodle^{TM} as the official platform, and additionally a group in Facebook^{TM} was opened.

Fifteen forums were designed corresponding to 15 topics of the Computer Programming subject. Each forum had two intentions: reflective and cognitive. Reflective intention was to promote expressing opinions about how students feel and for socialization, while the cognitive intention is for the students to solve problems related to the subject in teamwork, the forum was designed for them to share issues related to problem resolution, either providing clues of where to search information, expressing ideas or doubts and requesting or providing specific help (the detailed study is in [20]).

A total of 361 messages of Moodle and 42 from Facebook were obtained. Once messages were recovered from platforms, units of analysis were established. According to Rourke [21], this process is called unitizing and it consists on identifying the segments of the transcript that will be recorded and categorized. In this sense, each message was divided into phrases, that is, a *sentence level*, because it was observed that in each one of them it could be found evidence of different quadrants, both of cognitive presence and emotions. For each sentence, the corresponding content analysis was carried out following the methodology proposed in the Practical research model. In this sense, it was necessary to apply the same methodology for the content analysis of emotions as follows:

First, it was assigned a name to each quadrant according to the type of emotions that correspond to the sentence (see Fig. 4). This was done, in order to establish emotional categories for content analysis, that is, each quadrant is considered a category. Subsequently, indicators were designed for messages classification.

Each message was classified, by three expert teachers in the Computer Programming subject, in two ways. First, with the indicator that matched on of the categories. Then, using indicators designed for emotions classification.

4.1   Results

The content analysis of each message was linked to both cognitive presence and emotions. It was found that in general for reflective intention, positive emotions predominated, while evidence of cognitive content was scarce or null in some cases.

Which may be normal due to the type of questions asked, for example, a question in the first forum: "How do you feel at the beginning of the course?".

In Table 1 is presented an example with a brief description about how the analysis was carried out for each message, in the example the first forum is used. The first column contains a sequential sentence number; in this case there are only two sentences. The second column has the original message in Spanish and its translation into English is in the third column. The Cognitive Quadrant (CQ) in this case indicates that there was no evidence of cognitive presence (None). In Emotion Quadrant (EQ) in which the sentence was categorized is expressed in the last fifth column. In this example, the student first sentence state corresponds to the emotions in quadrant IV (e.g. excited or enthusiastic) and the second sentence, in row two, corresponds to quadrant I (e.g. hopeful, interest or confident).

Table 1. Description of sentence analysis for a reflexive question.

S	Spanish language	English language	CQ	EQ
1	Estoy entusiasmado por aprender C y me gusta bastante el hecho de que escribamos los programas en el cuaderno	I am excited to learn C and I quite like the fact that we write the code in the notebook	None	IV
2	Este método logra que me memorice la sintaxis de C y pienso que fortalecerá mis habilidades referentes a la lógica de programación	This method manages to memorize C's syntax and I think it will strengthen my skills regarding programming logic	None	I

In Table 2 a dialogue retrieved from a cognitive question is presented as an example. In this case, sentence 1 in second row contains the word "problem", which puts the student in the cognitive quadrant 1 and quadrant I of emotions since the interpretation could be interest, curiosity, or intrigue. Then, speech changes to cognitive quadrant 2, in sentences number 2, where the student proposes a solution and does it with confidence or as enlightenment, quadrant I emotional. Finally, the student shares his/her solution that takes him/her to cognitive quadrant 3 and to IV in emotions for satisfaction, enthusiastic or proud. But in that same message, sentence 3, he/she expresses curiosity or intrigue, which are in emotional quadrant I.

From six cognitive questions, in terms of emotion most of them converged in insight, enlightenment or epiphany, and most of them finished in "Exploration" regarding the Cognitive presence quadrant.

As forums go on, different cognitive an emotional states change. In Fig. 5 the dynamics that occurred with two students is presented. Student A, who participated more and was the most outstanding student of the course is compared with Student B, who had little participation and did not accredit the course.

Table 2. Description of sentence analysis for a cognitive question.

S	Spanish Language	English Language	CQ	EQ
1	Aquí el único problema que tengo encontrar una función de C que me convierta los caracteres a números enteros	Here the only problem I have is finding a C function that converts the characters to integers	1	I
2	Se puede pedir el número como string para ciclarlo más fácilmente	We can ask the number as a string to cycle it more easily	2	I
3	Al final lo resolví quitándole 48 al código ASCII de cada uno, pero me quedé con la duda de si había una forma más elegante de hacerlo	In the end I solved it by removing 48 of the ASCII code from each one, but I was left wondering if there was a more elegant way to do it	3	IV, I

The arrow indicates a transition in the cognitive state. The numbers in parentheses at beginning and end of the arrow indicate the cognitive and emotion quadrant classification. The number of the forum is displayed in the sequence arrow in the middle. When there was no participation of the student, there is a dash instead.

For example, in the second forum (F2), Student A gave evidence of "Exploration" or cognitive quadrant 2; and showed an emotion of *Learning Interesting* or quadrant 1 of emotions. Then his/her speech changed to quadrant 3 "Integration", as he/she continued with a positive emotion in quadrant I. As for student B, in the same forum (F), he/she was stuck in cognitive quadrant 2, with a positive emotion of quadrant I, which might be interpreted as the student was far from solving the problem.

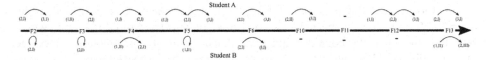

Fig. 5. Dynamics of affective-cognitive states in each forum.

It can be observed, Student B did not properly transit through the cognitive quadrants, which led him/her to land in the negative quadrants of emotions in forums 4, 5 and 13 (F4, F5 and F13). He/she managed to get out of the negativity state of mind in forum 4 (F4), which probably made him/her to move to the next cognitive quadrant. However, in F5 he/she was stuck in cognitive quadrant 1, perhaps due to negative emotion experienced. Finally, in F13 although he/she advanced in cognitive quadrants, he/she probably was lost due to confusion and frustration emotions. Whereas, Student A presents more transitions while his/her emotions were more stable.

To clarify, the presence of cognitive quadrant 4 corresponding to the "Resolution" category of cognitive process is almost null in the forums, because the products of the activities were presented in a face-to-face sessions. In this sense, the activities in discussion forums should be designed in a way that promotes dialogues of how the students managed themselves to solve the problem.

It is worth to mention that students who actively participated in most forums were the ones that reached the best course evaluation. Their transition through different cognitive quadrants, as well as emotions quadrants can be observed.

It is possible to detect in which quadrant of the cognitive process the students are and what emotions they are experiencing, at any moment. In such a way that if they are in a negative position with respect to a cognitive phase and there is no progress, the teacher might somehow intervene (Teaching presence).

Based on this study and with the aim of automating the whole process previously exposed, the following computational model is proposed.

5   Computational Model of Cognitive-Emotional Process for Engineering Students

A cognitive-emotional model should provide relevant information respect to certain level of cognitive presence, as well as the emotions that can project or inhibit the students learning process. This permits the teacher to monitor and more effectively achieve effective learning focusing on those students who need it most. Automatizing this model will provide the possibility of giving feedback immediately, personally and directly to the teacher and/or the students.

Our model is composed of three modules: Preprocessing, Sentiment-Cognitive Analysis, and Inference module, and additional components. Figure 6 shows these components, next explained.

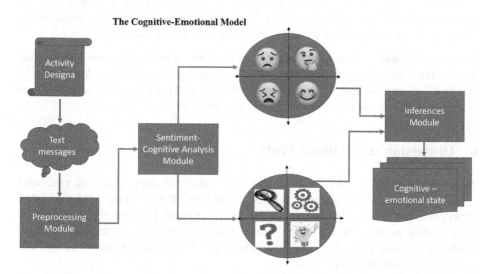

Fig. 6. Computational model of cognitive emotional process for engineering students.

Activity Design Component. The activity design is a fundamental part of this model; a good implementation will highlight the expected dialogues. If we intend to obtain meaningful information from both the cognitive process and the emotions, the activities should encourage these key phases.

Messages Recovery Component. The computer model starts by recovering dialogues that in the CMC media, which generates unstructured text files that must be prepared in Preprocessing Module.

Preprocessing Module. This is the first module; it will be responsible for outputting the adjusted text file, ready to be processed. Here important issues have to be considered about how it is convenient to divide the message and what grammatical elements are required; for this there are different preprocessing types, for example, tokenization, lemmatization, part of speech, among others. In the literature two tools stand out in the Sentiment Analysis area. NLTK [22] the natural language toolkit that is one of the best known and most used natural language process (NLP) libraries in the Python ecosystem, and Freeling [23], a C++ library that provides language analysis functionalities for several languages including Spanish. Although preprocessing is a stage of any SA system, it was set apart for a better understanding.

Sentiment-Cognitive Analysis Module. This module will be responsible for labeling each message with its corresponding cognitive and emotional category. The main computational approaches for this task are Machine Learning, Lexicon based, and Hybrid systems. A hybrid approach is the advisable approach given the complexity of the task; it has to be kept in mind that two types of classification will be done. A Systematic Literature Review for Sentiment Analysis for Spanish Language was carried out [24] following the methodology proposed in [25]. It was found that regarding machine learning algorithms, Vector Support Machine is the most widely used. Also, some lexical resources are available for Spanish language.

Inferences Module. Once the message is labeled, the inferences module will evaluate the cognitive and emotions quadrants, which can generate information in real time for teachers to detect students with problems of learning and the subjects in which they are failing in order to make decisions on how to help them.

6   Discussion and Future Work

An exploratory study was conducted to understand the implications of evaluating cognitive and emotional elements for SMET students. Results were used to create a conceptual model from the adapted models of The Community Inquiry Model [9] for the cognitive aspect, and for the emotional aspect the Model of interaction between emotions learning [19]. A proposal to automatize the conceptual model is also discussed into a computational model.

The computational model contains three modules: Preprocessing, Sentiment-Cognitive Analysis and Inference modules to label text sentences on cognitive and emotional states to make inferences regarding the students' learning process, using mainly NLP and Machine learning techniques.

Our future work consists of developing the computational model that automates this whole process, which evaluates the validity of our proposal through experimental scenarios of collaborative learning.

References

1. Goleman, D.: Emotional Intelligence. Bantam Books Inc., New York (1995)
2. Rosalind, W.P.: Affective Computing. First MIT, Cambridge (2000)
3. Kort, B., Reilly, R., Picard, R.W.: An affective model of interplay between emotions and learning: reengineering educational pedagogy-building a learning companion. In: 2001 Proceedings of the IEEE International Conference on Advanced Learning Technologies, pp. 43–46 (2001)
4. Seat, E., Parsons, J.R., Poppen, W.A.: Enabling engineering performance skills: a program to teach communication, leadership, and teamwork. J. Eng. Educ. **90**(1), 7–12 (2001)
5. Tromp, J.G., Steed, A., Wilson, J.R.: Systematic usability evaluation and design issues for collaborative virtual environments. Presence Teleoperators Virtual Environ. **12**(3), 241–267 (2003)
6. Poria, S., Cambria, E., Bajpai, R., Hussain, Λ.: Λ review of affective computing: from unimodal analysis to multimodal fusion. Inf. Fusion **37**, 98–125 (2017)
7. Garrison, D.R., Cleveland-Innes, M.: Facilitating cognitive presence in online learning: interaction is not enough. Am. J. Distance Educ. **19**(3), 133–148 (2005)
8. Henri, F.: Computer conferencing and content analysis. In: Kaye, A.R. (ed.) Collaborative Learning Through Computer Conferencing. NATO ASI Series (Series F: Computer and Systems Sciences), vol. 90. Springer (1992)
9. Garrison, D.R., Anderson, T., Archer, W.: Critical inquiry in a text-based environment: computer conferencing in higher education. Internet High Educ. **2**(2–3), 87–105 (1999)
10. Marcelo García, C., Perera Rodríguez, V.H.: Comunicación y aprendizaje electrónico:la interacción didáctica en los nuevos espacios virtuales de aprendizaje. Rev. en Educ. **343**, 381–429 (2007)
11. Rourke, L., Anderson, T.: Using web-based, group communication systems to support case study learning at a distance. Int. Rev. Res. Open Distrib. Learn. **3**(2), 1–13 (2002)
12. Garrison, D.R., Anderson, T., Archer, W.: Critical thinking, cognitive presence, and computer conferencing in distance education. Am. J. Distance Educ. **15**(1), 7–23 (2001)
13. Plutchik, R., Kellerman, H.: Theories of Emotion, of Emotion: Theory, Research, and Experience, vol. 1 (1980)
14. Russell, J.A.: A circumplex model of affect. J. Pers. Soc. Psychol. **39**(6), 1161–1178 (1980)
15. Ekman, P.: Universal facial expressions of emotion. Calif. Ment. Heal. Res. Dig. **8**(4), 151–158 (1970)
16. Ortigosa, A., Martí-n, J.M., Carro, R.M.: Sentiment analysis in Facebook and its application to e-learning. Comput. Human Behav. **31**, 527–541 (2014)
17. Arguedas, M., Xhafa, F., Casillas, L., Daradoumis, T., Peña, A., Caballé, S.: A model for providing emotion awareness and feedback using fuzzy logic in online learning. Soft. Comput. **22**(3), 963–977 (2018)
18. Cabada, R.Z., Estrada, M.L.B., Hernández, F.G., Bustillos, R.O., Reyes-Garcí-a, C.A.: An affective and Web 3.0-based learning environment for a programming language. Telematics Inf. **35**(3), 611–628 (2018)
19. Kort, B., Reilly, R.: Theories for deep change in affect-sensitive cognitive machines: a constructivist model. J. Educ. Technol. Soc. **5**(4), 56–63 (2002)

20. Osorio Angel, S., Peña Pérez Negrón, A., Estrada Valdez, A.: Online discussion forums: a didactic strategy to promote collaborative learning in computer programming subject., No Publ
21. Rourke, L., Anderson, T., Garrison, D.R., Archer, W.: Methodological issues in the content analysis of computer conference transcripts. Int. J. Artif. Intell. Educ. **12**, 8–22 (2001)
22. Bird, S., Klein, E., Loper, E.: Natural Language Processing with Python: Analyzing Text with the Natural Language Toolkit. O'Reilly Media, Inc., (2009)
23. Padró, L., Stanilovsky, E.: Freeling 3.0: Towards wider multilinguality. In: LREC2012 (2012)
24. Osorio Angel, S., Peña Pérez Negrón, A., Estrada Valdez, A.: Sentiment analysis in text for affective computing: a survey in Spanish language., No Publ
25. Brereton, P., Kitchenham, B.A., Budgen, D., Turner, M., Khalil, M.: Lessons from applying the systematic literature review process within the software engineering domain. J. Syst. Softw. **80**(4), 571–583 (2007)

Algorithm Proposal to Control a Robotic Arm for Physically Disable People Using the LCD Touch Screen

Yadira Quiñonez[1]([⊠]), Oscar Zatarain[1], Carmen Lizarraga[1],
Juan Peraza[1], and Jezreel Mejía[2]

[1] Universidad Autónoma de Sinaloa, Mazatlán, Mexico
{yadiraqui, carmen.lizarraga, jfperaza}@uas.edu.mx,
2016030617@upsin.edu.mx
[2] Centro de Investigación En Matemáticas, Zacatecas, Mexico
jmejia@cimat.mx

Abstract. This research focuses on people who have an issue to move their bodies or do not have enough force to move it, with the purpose to help in their control their moves to reach objects in their quotidian live with a fast process, in this sense, it has been conducted different interviews with persons that have coordination problem in their bodies or could have them. Therefore, in this work, it is proposed an algorithm to create a touch control for people that can use to manipulate a robotic arm with only one finger, faster with less effort. The Algorithm is based on two discrete functions that are characterized by depending on the problem that is going to work and get the most efficient result. These results were tested in Arduino and LCD-touch with a digital and discrete treatment to finding the shortest and effective way of acting.

Keywords: Robotics arm · Assistive robotic manipulators · Assistive technology · Physically disabled people · LCD touch screen

1 Introduction

The applications of robotic arms have been developing constantly and their process is each time more efficient, since the concept "Robota" of Karel Capek and industrial robotic arm of George Devol until the evolution of their uses and the applications of different areas just like the industry, science, entertainment, medicine, and rehabilitation. At the same time, digital and analogic electronic has been developed with the purpose to make current difficult activities and many sciences and engineering areas have given tracing such as digital control, analogic control, modern control, and other engineering areas to make more efficient the manipulation. Although robotic arms are fast when the user programming it in an industry, those robots follow one instruction and repeat again, nevertheless, that does not happen when a person control a robotic arm, sometimes the robotic arm could be very sophisticated but its control very slow and very trouble so much so that they need a lot of effort and along wait.

J. Mejía et al. (Eds.): CIMPS 2019, AISC 1071, pp. 187–207, 2020.
https://doi.org/10.1007/978-3-030-33547-2_15

In the last two decades, automation and control have become a topic of interest for researchers from different areas. Mainly, in the industrial robotics [1, 2] and in the robotic systems applied in the medical area, such as tele-operated surgery [3], surgical pattern cutting [4], according to Shademan et al. the precision of an industrial robot offers great advantages in this area [5]. Currently, there are endless technological developments and various applications such as prostheses [6], orthoses [7], exoskeletons [8–10], and devices for tele-operation in order to improve human capabilities [11–14]. According to Chung et al. conducted a review of the literature about the different robotic assistance manipulators from 1970 to 2012, mentioning that reliability, profitability, appearance, functionality, and ease of use are determining factors to achieve successful commercialization [15]. Some related works for the rehabilitation of patients with reduced movement capacity have used virtual reality to improve the manipulation of movements and positively influence a virtual environment [16–19]. Recently, in a work by Atre et al. have used techniques of computer vision to create a robotic arm controlled by gestures, they mention the importance that a robotic arm is more efficient in relation to accuracy and speed [20], in another works, Badrinath et al. propose the control of a robotic arm using a Kinect 3-D camera to track the movements of people through computer vision technique [21].

2 Algorithm Proposal

Nowadays, it is very known the different uses of the joystick in the manipulation of the robot arm in many industries, videogame and even many people use a robot arm which is controlled with Joystick [22–24]. Nevertheless, also it is very known that persons use a lot of electronic devices which are controlled with touch and each time, people are more familiarize with a touch screen, therefore, this control is based on an LCD Touch Screen due to nowadays the touch screens are very used and could be more comfortable. Moreover, it does not need much energy to interact with a touch screen while it is used, and it is another reason which is considerate for its implementation in this investigation.

According to the different necessities or problems of physically disabled people, the LCD Touch can help to cover all the necessities in the best way, because of it is important that people use only one finger to control the robotic arm for a comfortable manipulation and with the fastest way, sometimes, the physical disable people can not use a lot of energy on how to push a button or move a joystick in extreme cases. In addition, it would be better to have a control to manipulate the robotic arm with less effort as possible as can be, due to that some emergency events could be very difficult to keep a force in the fingers to move the robotic arm and waiting for that it can reaches to a certain point on space. Also, this algorithm that is implemented on the LCD Touch Screen, it has been thought for all people that can have a physically disabled, such as people that are sick of diabetes and have hypoglycemia or people who suffer of lupus, then, they just have to touch a point on the LCD Touch Screen to move a robotic arm in emergency cases, and it is very important to mention too that people with lupus or diabetes can have confusion during a hypoglycemia (in the case of diabetic people) or another emergency case, and be able to move simply touching a coordinate of the LCD touch screen can save lives.

This Algorithm can reduce the move time because it is based on the choice of the way with least time as possible as can be and at the moment that it is had a touch point on the LCD Touch Screen, they are had coordinates on axis X and Y, therefore, it can be moved many motors at the same time which is the fastest way to reach a certain point in the space and the best representation of the human arm too. Unlike with joystick control, only it can be moved one motor per move, always the move time is linear and worth mentioning that the patients need to apply a constant effort and movement to get at a specific point on the space while that with an LCD Touch Screen it is necessary to touch one time a coordinate and the robotic arm will get to position desired. Another characteristic of this algorithm is the way to use this touch control because is very natural because the algorithm is based on the movement of a finger, a movement so natural as it can be and be a machined movement that does not need a lot of effort and concentration unlike of the methods which control a robotic arm with the thought sensors.

In this sense, it proposes an algorithm, which provides a fast movement with less effort, is proved in Arduino with an LCD Touch and the user touches only one time the coordinate in the LCD to move the robotic arm in a specific environment point and workspace of the robotic arm. Besides, its function is based on two discrete functions that are characterized by depending on the problem that is going to work and get the most efficient result.

3 Algorithm Conditions

The strategy to describe the robot arm moves with a success for us when is almost similar to the human arm, and many systems try to describe it. This algorithm tries to describe the human arm moves and concentrated in moved the motors at the same time to get a specific point of the environment space. With an LCD and touching a coordinate in the space of LCD, it will know a point on the axis x and y, therefore, one motor will need to rotate some degrees to get x and y coordinates that have been touched. So, the robotic arm as similar moves as a human arm and it moves at the three axes in the environment space, then, the algorithm needs the two coordinates and consequently, division the two numbers to place them in the domain of the trigonometric function. The first condition will be that the coordinates of x and y are two hicks of a triangle, therefore, the algorithm works with a tangent function, as a result, the degree α will be calculated. Figure 1 shows an example of touch point in the LCD Touch Screen and representation of the variable's x, y and α in the space work to calculate α and symbolize the first condition.

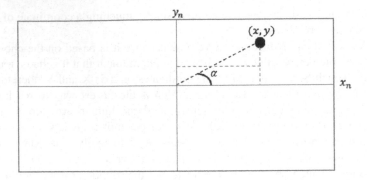

Fig. 1. Touch point in the LCD-Touch screen with representation of the variables x, y and α.

Therefore, α is defined such as:

$$\alpha = \text{tang}\left(\frac{y}{x}\right) \tag{1}$$

where x and y are real numbers and $x > 0$.

3.1   First Condition

When a certain touch point is located for a beta less or equal than 45°, the first condition is fulfilled. Figure 2 shows a representation of the first condition on the triangle's area which the finger is touching.

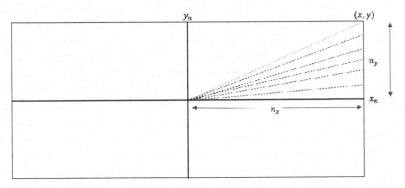

Fig. 2. Triangle's area where the first condition is fulfilled for α less or equal than 45°.

The number n_x and n_y are whole numbers which represent the amount of pulses that x and y need to get the touchpoint. x_n and y_n are the limit numbers that represent the space work of LCD Touch and (x, y) is the coordinate that has been touched. Therefore, when α is less or equal that 45°, then:

$$\alpha \le 45° \Rightarrow \tan^{-1}\left(\left(\frac{y}{x}\right)\right) \le 45 \Rightarrow \frac{y}{x} \le \tan(45) \Rightarrow \frac{y}{x} \le 1 \tag{2}$$

therefore, $y \le x$ this means that $n_y \le n_x$. For programming, pulses conditions of axis x and y for when beta is less or equal than 45°: *if* $\alpha = 45$ *then* $n_x = n_y$ and *if* $\alpha < 45$ *then* $n_x > n_y$.

When α is equal to 45°, this means that:

$$\lim_{\left(\frac{y}{x}\right)\to 1}\alpha = \lim_{\left(\frac{y}{x}\right)\to 1}\tan^{-1}\left(\frac{y}{x}\right) = \tan^{-1}(1) = 45 \tag{3}$$

However, the lateral limits are equal when $\left(\frac{y}{x}\right)$ is approximate to 1 and $\left(\frac{y}{x}\right)$ is approximately by right, $\left(\frac{y}{x}\right)$ does not meet with the first condition, and in this case, y is greater than. Then, it is necessary to propose the second condition.

3.2   Second Condition

The second condition is presented when x is fewer than y and this mean that $\left(\frac{y}{x}\right)$ is going to be greater than one, in this case, α is greater or equal than 45° and for a very large $\left(\frac{y}{x}\right)$, the maximum amount of α is 90°, then, for the second condition, it proposes that α is:

$$45 \le \alpha \le 90 \tag{4}$$

such that $y \ge x$ and consequently $\left(\frac{y}{x}\right) \ge 1$. The second condition is represented in Fig. 3 which shows the triangle's area where the touch point is located:

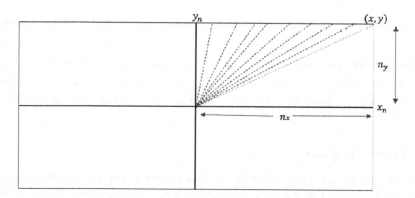

Fig. 3. Location of the touch point on triangle's area where the second condition is fulfilled.

Then, the rules for programming are: *if* $\alpha = 45 \Rightarrow n_x = n_y$ and *if* $\alpha > 45$ *and* $< 90 \Rightarrow n_x < n_y$. Pulses conditions of axis "x" and "y" for when beta is greater or equal than 45° and less or equal than 90°.

3.3   Third Condition

It knows that the limit of tangent when $\left(\frac{y}{x}\right)$ tends to a great number (that is mean when the limit tends to infinite), always the limit will be 90°, therefore, in the third condition, it is going to introduce other rules to reach 135°.

$$\lim_{\left(\frac{y}{-x}\right)\to -\infty} \alpha + 180 = \tan^{-1}(-\infty) + 180 = -90 + 180 = 90 \qquad (5)$$

In the third condition, always x will be a negative number, that is why in the condition always will a sum of 180° and the result of α will be between 90 and 45° to reach 135°.

$$90 \leq \alpha + 180 \leq 135 \qquad (6)$$

Such that $y > 0 > -x$. Therefore, the third condition is represented as in Fig. 4:

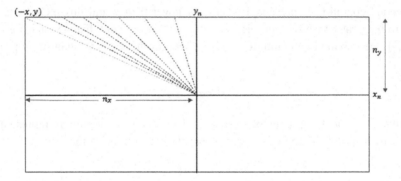

Fig. 4. Location of the touch point on triangle's area where the third condition is fulfilled.

The rules for programming are: *if $\alpha = 45 \Rightarrow tan^{-1}\left(\frac{y}{-x}\right) + 180\,such\,as\,x = y\,and$* $n_x = n_y$ *and if $90 < \alpha < 135 \Rightarrow tan^{-1}\left(\frac{y}{-x}\right) + 180\,such\,as\,x = y\,and\,n_x < n_y$*. Pulses conditions of axis x and y for when beta summing 180° is greater or equal than 90° and less or equal than 135°.

3.4   Fourth Condition

The fourth condition is very similar to the third condition, nevertheless, the variables x and y change. In this condition x will be greater than y to reach 180°, therefore, in this condition α is between 135 and 180°.

$$135 \leq \alpha + 180 \leq 180 \qquad (7)$$

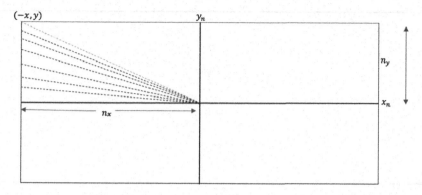

Fig. 5. Location of the touch point on triangle's area where the fourth condition is fulfilled.

Such that $x \geq y$ and $\left(\frac{y}{-x}\right) \leq 0$. The fourth condition is represented in Fig. 5, which represent the location of the touch point when the variable α with the sum of 180° is greater or qual than 135° and less or equal than 180:

The rules for programming are: *if* $\alpha + 225 = 180 \Rightarrow tan^{-1}\left(\frac{y}{-x}\right) + 225$ *such as* $x = y$ *and* $n_x = n_y$, *if* $\alpha + 180 = 180 \Rightarrow tan^{-1}\left(\frac{y}{-x}\right) + 180$ *such as* $y = 0$ *and* $n_y = 0$, *if* $135 < \alpha < 180 \Rightarrow tan^{-1}\left(\frac{y}{-x}\right) + 180$ *such as* $x > y$ *and* $n_x > n_y$. Pulses conditions of axis x and y for when beta summing 180° is greater or equal than 135° and less or equal than 180°. This condition must add another requirement, which is the first rule, due to it has to meet the representation of Fig. 5 when y as equal as x.

4 Analysis

A NEMA motor of 1.8° per step is used, when the touch point is touched in one of the four condition, then, the LCD gives two different coordinate which is (x, y) and then, α is calculated depending of which of the four conditions is touching. Therefore, α needs a function which calculated the pulses for reach the coordinate (x, y) that the finger has touched. A function is defined as:

$$f : n \rightarrow \alpha, f(n) = 1.8n : 0 \leq n \leq 100 \, n \in \mathbb{Z} \qquad (8)$$

4.1 Maximum Travel Time of α

The motor α always is located in 0° as shown in Fig. 6:

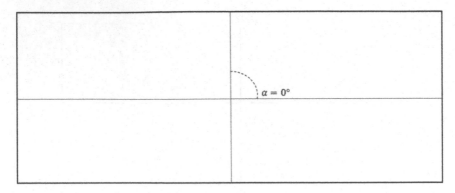

Fig. 6. Representation of α motor for when it starts to work.

If α has to arrive to 180° and α star with 0°, then, the move motor is given for 100 pulses for that α gets 180°, so that:

$$f(n) = 1.8n, \text{ if } n = 100 \text{ then} f(100) = 1.8(100) = 180° \tag{9}$$

Also, the width of each pulse is 10 ms and each pulse, α gets a step of 1.8°. Theoretically, α needs a time of 1 s to get 180°. Now, $f(n) = 1.8n$ work well, nevertheless, in this subject is going to include a new function, which is faster and accurate on some occasions:

$$f(n) = \sum_{n=0}^{s} 1.8n \tag{10}$$

such that $s = \{1, 2, 3, 4, \ldots, 13\}$. In this case, s is a finite set that has the numbers of pulses for $f(n)$ needs. Now, these two functions are translated so that Arduinos can read them. Then, the process is constructed as follows in Fig. 7, which is a diagram that describe the process for obtaining the α Motor's movements:

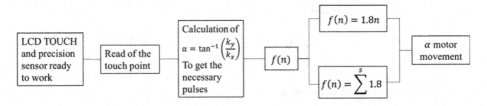

Fig. 7. Diagram in which the process for obtaining the α Motor grades is represented

5 Algorithm Function

Now that all the conditions are met, the function of the algorithm can be explained. It is know that the $f(n') = 1.8n'$ need more pulses than $f(n) = \sum_{n=0}^{s} 1.8n$, because of

$f(n) = \sum_{n=0}^{s} 1.8n$ just needs 13 pulses to reach $180°$ and $f(n') = 1.8n'$ needs 100 pulses to reach $180°$, due to the different work of each pulse of $f(n')$ and $f(n)$. The pulses of $f(n) = \sum_{n=0}^{s} 1.8n$ works by increasing the width of the pulses, this follows:

$$f(n) = \sum_{n=0}^{s} 1.8n = 1.8(1) + 1.8(1 + 1) + 1.8(1 + 1 + 1) + \dots + 1.8(s)$$
$$= 1.8(1) + 1.8(2) + 1.8(3) + \dots + 1.8(s)$$

(11)

As a pulse has 10 ms of width, then, the width of the pulses will increment 10 ms each time n increase until reach s. Table 1 explains the width increment pulses of each pulse of $f(n)$.

Table 1. Pulses of function $f(n)$ and width in time of each n pulse.

$f(n)$ Pulses n	Time
1	10 ms
2	20 ms
3	30 ms
4	40 ms
5	50 ms
6	60 ms
7	70 ms
8	80 ms
9	90 ms
10	100 m
11	110 ms
12	120 ms
13	130 ms

5.1   Algorithm Representation

This process is focused on the algorithm, which tries to find the nearest way to reach an objective. This algorithm works with two options which there can be mixed and if the target is not reached, the ways to get the target will keep increasing. The options of this algorithm increasing of 2^n form and it chooses the fastest way, such as it is shown below in Fig. 8:

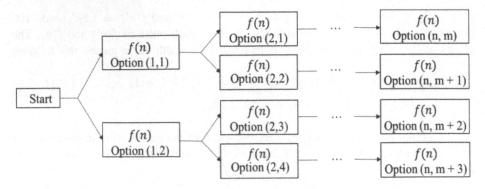

Fig. 8. Algorithm representation of 2^n form

According to the diagram, the algorithm has 2^n forms for select the ways and if it wants to know how many forms it is obtained in a specific time, then, this algorithm could be expressed as the next way:

$$a \geq \lim_{t_s \to b} \frac{2^n}{t_s} \geq a - b \tag{12}$$

where a : desired number and 2^n : number of ways:.

This in equation always must meet with the condition bellow:

$$a < 2^n \text{ and } b \leq \frac{2^n}{a} \tag{13}$$

Then, when b is known, it can be equaled to the ratio:

$$b = \left(t_p/t_d\right) \tag{14}$$

where t_d is the limit inferior time and t_p is the limit superior time

By last, the time that is sought is found between t_d and t_p, it mean

$$t_p \geq t \geq t_d \tag{15}$$

where t is the time sought.

Other condition which the inequation must meet, there are:

First, it must be treated that the limits are whole, therefore, it always must be round. Also, if the unit is greater than e it is necessary to move on to the next ten and if the units are less than e it is necessary to sum e. For this case $e = 5$. For example: if the desired number is $3.6°$ $(a = 3.6°)$, using the in Eq. (12), then:

$$3.6 \geq \lim_{t_s \to b} \frac{2^n}{t_s} \geq 3.6 - b \tag{16}$$

Also, in this case $n = 2$ due to $2^n = 2^2 > 3.6$. Then, $b \leq \frac{4}{3.6}$, therefore $b \leq 1.\overline{11}$

$$3.6 \geq \lim_{t_s \to b} \frac{2^n}{t_s} \geq 2.4889 \tag{17}$$

For this reason and following the equalization, then

$$1.\overline{11} \geq (t_p/t_d) \tag{18}$$

How $n = 2$, then, the inferior limit is 20 ms:

$$1.\overline{11} \geq \left(\frac{t_p}{20 \text{ ms}}\right), \text{ therefore, } t_p = 22.\overline{22} \tag{19}$$

According to the conditions, $t_p = 27$, then, the time which is has been sought (t), it is between:

$$27 \text{ ms} \geq t \geq 20 \text{ ms} \tag{20}$$

In conclusion, there are 4 ways for a time that is between 27 ms and 20 ms.

5.2 Simulation of Arduino Code

While it was explained about the different works of each function in the algorithm function part, all of that characteristics land of the functioning and the form of how the algorithm will be code. Now, it is known that each function has a different pulses width range and join together to make another function that has the combination pulses of the functions $f(n')$ and $f(n)$. First, the ports are defined in the Arduino where the pulses of $f(n')$ and will be throw $f(n)$. In this case, it will be:

```
#define Sum 50
#define nn 52
```

Where #define Sum 50 is the output of pulses of $f(n) = \sum_{n=0}^{s} 1.8n$ and #define nn 52 is the output of $f(n') = 1.8n'$. After, it is necessary getting the desired degrees of, which are found them touching the LCD Touch screen, the following code example shows the parameters of touch point on the LCD Touch screen and the outputs that are necessary to this touch point.

```
TSPoint p = ts.getPoint();
  pinMode(XM, OUTPUT);
  pinMode(YP, OUTPUT);
  if (p.z > MINPRESSURE && p.z < MAXPRESSURE) {
    p.x = map(p.x, TS_MINX, TS_MAXX, tft.width(), 0);
    p.y = map(p.y, TS_MINY, TS_MAXY, tft.height(), 0);
    Serial.print("("); Serial.print(p.x);
    Serial.print(", "); Serial.print(p.y);
    Serial.println(")");
    if (p.x > 9 && p.x < 21) {
      if (p.y > 20 && p.y <150) {
        for(i=0;i<1;i++)
          {
              digitalWrite(Sum, HIGH);
              tft.fillCircle(120, 160, 50, GREEN);
              tft.setCursor(95, 150);
              tft.println("LED");
              delay(10);

              digitalWrite(Sum, LOW);
              tft.fillCircle(120, 160, 50, RED);
              tft.setCursor(95, 150);
              tft.println("LED");
              delay(0);

              digitalWrite(Sum, HIGH);
              tft.fillCircle(120, 160, 50, GREEN);
              tft.setCursor(95, 150);
              tft.println("LED");
              delay(20);

              digitalWrite(Sum, LOW);
              tft.fillCircle(120, 160, 50, RED);
              tft.setCursor(95, 150);
              tft.println("LED");
              delay(0);
          }
      }
    }
```

In this case, it is had the pulse of $f(n)$ when n is equal to two, then it is had a sort of pulse such as it is shown in Fig. 9.

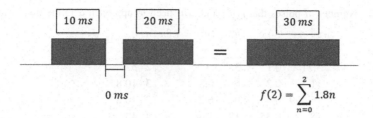

Fig. 9. Example of answer pulse of the code for when α is approximated to 5.71° and n = 2.

In the time that the degrees are augmented, the pulses of the function $f(n)$ and the output of this function (#define Sum 50) will have a greater width, such as it is shown in Fig. 10.

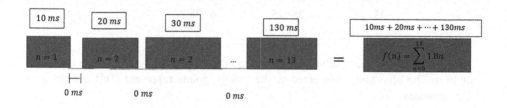

Fig. 10. Representation of how the function $f(n)$ is working through that α is growing to reach until 163.8°.

6   Results

The moves in the axis of ordinates of the robot are the degrees that be obtained from the location of the finger in pressure sensor and degrees of the axis of abscissas of the robot is the ratio (y/x) with the pulses (n) that are needed to reach the degrees desired of the functions $f(n') = 1.8n'f(n') = 1.8n'$ and $f(n) = \sum_{n=0}^{s} 1.8n$. According to Table 2, it shows the different ranges of each function that are necessary to reach each degree amount for when α is greater or equal than 0 and less or equal than 45°:

Table 2. Number of pulses that $f(n')$ and $f(n)$ need to approach α in the first condition.

α	$\left(\frac{y}{x}\right)$	n'	$f(n') = 1.8n$	n	$f(n) = \sum_{n=0}^{s} 1.8n$
5.71	0.1	3	5.4	2	5.4
11.3	0.2	6	10.8	3	10.8
16.699	0.3	9	16.2	3	10.8
18	0.35	12	21.6	4	18
21.8	0.4	12	21.6	4	18
26.56	0.5	14	25.2	4	18
27	0.5095	14	25.2	5	27
30.96	0.6	14	25.2	5	27
34.99	0.7	17	30.6	5	27
37.8	0.7756	19	34.2	6	38.7
38.65	0.8	21	37.8	6	38.7
41.98	0.9	23	41.4	6	38.7
45	1	25	45	6	38.7

Figure 11 shows the results of each function for when the functions are precise or near to α. The blue line represented α, the orange points represent $f(n')$ and the gray points represent $f(n)$.

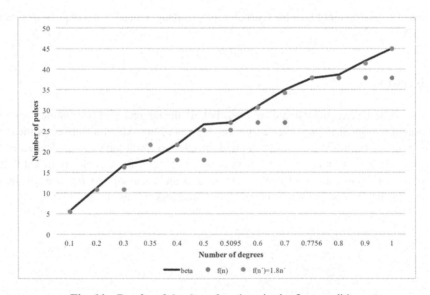

Fig. 11. Results of the three functions in the first condition.

Then, it has a new function as:

$$f(n) := \begin{cases} \sum_{n=0}^{s} 1.8n \; if \; 0 \leq \left(\frac{k_y}{k_x}\right) \leq 0.35, \; whits = \{1,2,3,4\} or \; if \; 0.51 \leq \left(\frac{y}{x}\right) \leq 0.79, \; s = \{5,6\} \\ 1.8n \; if \; 0.4 \leq \left(\frac{y}{x}\right) \leq 0.5 \; or \; if \; 0.6 \leq \left(\frac{y}{x}\right) \leq 0.7 \\ 1.8n \; if \; 0.8 \leq \left(\frac{x}{y}\right) \leq 1 \end{cases}$$

$$(21)$$

When α is greater that 45 and less or equal than 90°, Table 3 indicate the ranges of each function to approach to α:

Table 3. Number of pulses that $f(n')$ and $f(n)$ need to approach α in the second condition

α	$\left(\frac{y}{x}\right)$	n'	$f(n') = 1.8n$	n	$f(n) = \sum_{n=0}^{s} 1.8n$
45	1	25	45	8	64.8
63.434	2	35	63	8	64.8
75.963	4	42	75.6	8	64.8
80.53	6	44	79.2	8	64.8
81	6.313	45	81	9	81
82.87	8	46	82.8	9	81
84.298	10	47	84.6	9	81
87.137	20	48	86.4	9	81
88.56	40	49	88.2	9	81
88.85	50	49	88.2	9	81
89.427	100	49	88.2	9	81
89.93	900	49	88.2	9	81

In Fig. 12 shows the results of each function for when the functions are precise or near to α. The functions $f(n')$ and $f(n)$ are very approach to α, although only two times $f(n)$ touches to α.
Then, the function result on the second condition is:

$$f(n) \begin{cases} 1.8n \; if \; 1 \leq \left(\frac{y}{x}\right) \leq 1.99 \\ \sum_{n=0}^{s} 1.8n \; if \; 2 \leq \left(\frac{y}{x}\right) \leq 2.2 \; or \; if \; 6 \leq \left(\frac{y}{x}\right) \leq 8 \\ 1.8n \; if \; 2.3 \leq \left(\frac{y}{x}\right) \leq 5.9 \; or \; if \; 8.1 \leq \left(\frac{y}{x}\right) \leq 1000 \end{cases}$$

$$(22)$$

When α is greater than 90° and less or equal than 135° (third condition). In this case, it has to mix the function $f(n')$ and $f(n)$ as a sum. Such as it shows in Table 4.

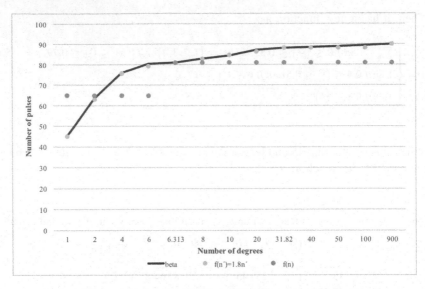

Fig. 12. Graphic results of $f(n')$ and $f(n)$ in the second condition.

Table 4. Number of pulses that $f(n')$ and $f(n)$ need to approach α in the third condition.

α	$\left(\frac{y}{x}\right)$	n'	$f(n') = 1.8n$	(n, n')	$f(n'') = f(n) + f(n')$
91.145	−50	50	90	(9, 0)	81
91.6	−30	51	91.8	(9, 0)	81
92.29	−25	51	91.8	(9, 0)	81
92.86	−20	52	93.6	(9, 0)	81
93.814	−15	52	93.6	(9, 0)	81
95.71	−10	53	95.4	(9, 0)	81
96.34	−9	54	97.2	(10, 0)	99
97.125	−8	54	97.2	(10, 0)	99
98.13	−7	55	99	(10, 0)	99
99	−6.313	55	99	(10, 0)	99
99.462	−6	56	100.8	(10, 1)	100.8
100.8	−5.242	56	100.8	(10, 1)	100.8
101.309	−5	56	100.8	(10, 1)	100.8
102.6	−4.0473	57	102.6	(10, 2)	102.6
104.036	−4	58	104.4	(10, 3)	104.4
108	−3.0077	60	108	(10, 5)	108
108.434	−3	60	108	(10, 5)	108
115.2	−2.125	64	115.2	(10, 9)	115.2
116.565	−2	65	117	(10, 10)	117
133.2	−1.064	74	133.2	(10, 19)	133.2
135	−1	75	135	(10,20)	135

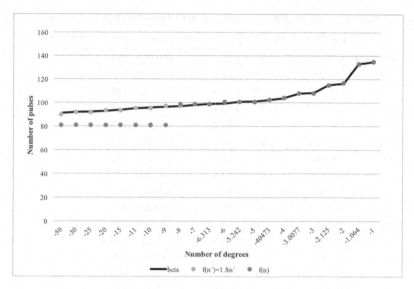

Fig. 13. Graphic results of $f(n')$ and $f(n)$ in the third condition.

The results of the function which is mixed with $f(n')$ and $f(n)$ is very precise and faster than $f(n')$. Figure 13 shows the results of each function and the approximate of function $f(n)$ and $f(n')$. The blue line represented $\propto$, the orange points represent $f(n')$ and the gray points represent $f(n)$ mix with $f(n')$.

Then, the function of programming is:

$$f(n) \begin{cases} 1.8n \ if -1000 \le \left(\frac{y}{x}\right) < -7 \\ \sum_{n=0}^{s} 1.8n \ if -7 \le \left(\frac{y}{x}\right) \le -6 \ or \ if \left(\frac{k_y}{k_x}\right) = -1 \\ \sum_{n=0}^{s} 1.8n + 1.8n' \ if \ 6 < \left(\frac{y}{x}\right) < 1 \end{cases} \quad (23)$$

When α is greater than 135° and less or equal than 180, this case is very similar to the last one, $f(n')$ and $f(n)$ are mixed. Table 5 shows the results bellow:

Table 5. Number of pulses that $f(n')$ and $f(n)$ need to approach α in the fourth condition.

α	$\tan^{-1}\left(\frac{y}{x}\right)+180$	n'	$f(n')=1.8n$	(n,n')	$f(n'')=f(n)+f(n')$
−0.9999	135.01	75	135	(10,20)	135
−0.99	135.28	75	135	(10,20)	135
−0.9	138.012	77	138.6	(10,20)	135
−0.85	139.012	78	140.4	(10,20)	135
−0.8	141.34	79	142.2	(10,20)	135
−0.75	143.036	80	144	(10,20)	135
−0.7	145.007	81	145.8	(10,20)	135
−0.6	149.036	83	149.4	(10,20)	135
−0.5	153.43	85	153	(10,20)	135
−0.4	158.198	87	156.6	(10,20)	135
−0.3	163.3	90	162	(10,20)	135
−0.2905	163.8	91	163.8	(13,0)	168.8
−0.2	168.69	94	169.2	(13,3)	169.2
−0.158	171	95	171	(13,4)	171
−0.1	174.28	97	174.6	(13,6)	174.6
−0.031	178.2	99	178.2	(13,8)	178.2
−1	180	100	180	(13,9)	180

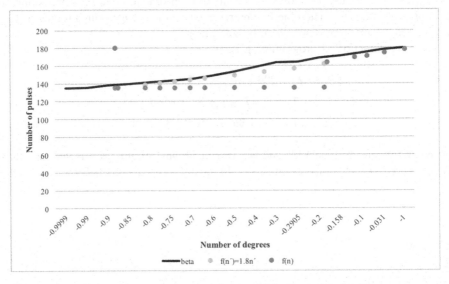

Fig. 14. Graphic results of $f(n')$ and $f(n)$ in the fourth condition.

The results of the function which is mixed with $f(n')$ and $f(n)$ is very precise and faster than $f(n')$ too. Figure 14 shows the results of each function and the approximate of function $f(n)$ and $f(n')$. Also, in this case $f(n')$ stays in 135 whiles (y/x) is less than −0.2905.

Then, the function of programming is:

$$f(n)\begin{cases} 1.8n \: if \: 0.999 \leq \left(\frac{y}{x}\right) < -0.3 \\ \sum\limits_{n=0}^{s} 1.8n \: if \: -0.29 \leq \left(\frac{y}{x}\right) \leq -0.25 \\ \sum\limits_{n=0}^{s} 1.8n + 1.8n'f \: -0.25 < \left(\frac{y}{x}\right) \leq -1 \: with \: s \: = \: 13 \: y \: 0 < n' \leq 9 \end{cases} \tag{24}$$

7 Conclusions

In this paper, an algorithm has been proposed to control a robotic arm for physically disabled people through an interface in the LCD touch screen which has been focused according to the needs of physically disabled people. This algorithm helps people to control a robotic arm with only one finger due to this applicable characteristic is the most important and it is based on two different discrete functions which were presented, one of the two functions is faster than the other one and the other function is more accurate (for this case), even though this case could be changed depending on the system which it is modeling with the shape 2^n with two different discrete functions, therefore, this algorithm could be developed in a future because of many technology process work with discrete functions such as in this case that has been studying. Also, the form of communicating us with the technology can be binary such as in the case of binary electronic, this is another reason the algorithm could be developed in a future job.

Another case, which is important to mention, is that this kind of algorithm can be used to people have faster tools, for this case, it is important that physically disabled people have a robotic arm that can move like a human arm with the lest effort as can be as possible. The algorithm can be programmed on Arduino with an easy way, therefore, the children and students can work with this algorithm and encourage the interest to create technology to help physically disabled people, in addition, this algorithm can be developed to create faster processes with an easy way and cheaper.

References

1. Grau, A., Indri, M., Bello, L.L., Sauter, T.: Industrial robotics in factory automation: from the early stage to the Internet of Things. In: 43rd Annual Conference of the IEEE Industrial Electronics Society, pp. 6159–6164. IEEE Press, Beijing (2017)
2. Yenorkar, R., Chaskar, U.M.: GUI based pick and place robotic arm for multipurpose industrial applications. In: Second International Conference on Intelligent Computing and Control Systems, pp. 200–203, Madurai (2018)
3. Burgner-Kahrs, J., Rucker, D.C., Choset, H.: Continuum robots for medical applications: a survey. IEEE Trans. Rob. 31(6), 1261–1280 (2015)

4. Murali, A., Sen, S., Kehoe, B., Garg, A., Mcfarland, S., Patil, S., Boyd, W.D., Lim, S., Abbeel, P., Goldberg, K.: Learning by observation for surgical subtasks: multilateral cutting of 3D viscoelastic and 2D orthotropic tissue phantoms. In: IEEE International Conference on Robotics and Automation, pp. 1202–1209. IEEE Press, Seattle (2015)
5. Shademan, A., Decker, R.S., Opfermann, J.D., Leonard, S., Krieger, A., Kim, P.C.: Supervised autonomous robotic soft tissue surgery. Sci. Trans. Med. **8**(337), 337ra64 (2016)
6. Allen, S.: New prostheses and orthoses step up their game: motorized knees, robotic hands, and exosuits mark advances in rehabilitation technology. IEEE Pulse **7**(3), 6–11 (2016)
7. Niyetkaliyev, A.S., Hussain, S., Ghayesh, M.H., Alici, G.: Review on design and control aspects of robotic shoulder rehabilitation orthoses. IEEE Trans. Hum. Mach. Syst. **47**(6), 1134–1145 (2017)
8. Proietti, T., Crocher, V., Roby-Brami, A., Jarrassé, N.: Upper-limb robotic exoskeletons for neurorehabilitation: a review on control strategies. IEEE Rev. Biom. Eng. **9**, 4–14 (2016)
9. Rehmat, N., Zuo, J., Meng, W., Liu, Q., Xie, S.Q., Liang, H.: Upper limb rehabilitation using robotic exoskeleton systems: a systematic review. Int. J. Int. Rob. App. **2**(3), 283–295 (2018)
10. Young, A.J., Ferris, D.P.: State of the art and future directions for lower limb robotic exoskeletons. IEEE Trans. Neural Syst. Rehabil. Eng. **25**(2), 171–182 (2017)
11. Makin, T., de Vignemont, F., Faisal, A.: Neurocognitive barriers to the embodiment of technology. Nat. Biomed. Eng. **1**(0014), 1–3 (2017)
12. Beckerle, P., Kõiva, R., Kirchner, E.A., Bekrater-Bodmann, R., Dosen, S., Christ, O., Abbink, D.A., Castellini, C., Lenggenhager, B.: Feel-good robotics: requirements on touch for embodiment in assistive robotics. Front. Neurorobot **12**, 1–84 (2018)
13. Jiang, H., Wachs, J.P., Duerstock, B.S.: Facilitated gesture recognition based interfaces for people with upper extremity physical impairments. In: Alvarez, L., Mejail, M., Gomez, L., Jacobo, J. (eds.) CIARP 2012. LNCS, vol. 7441, pp. 228–235, Springer, Heidelberg (2012)
14. Kruthika, K., Kumar, B.M.K., Lakshminarayanan, S.: Design and development of a robotic arm. In: IEEE International Conference on Circuits, Controls, Communications and Computing, pp. 1—4. IEEE Press, Bangalore (2016)
15. Chung, C.S., Wang, H., Cooper, R.A.: Functional assessment and performance evaluation for assistive robotic manipulators: literature review. J. Spinal Cord Med. **36**(4), 273–289 (2013)
16. Perez-Marcos, D., Chevalley, O., Schmidlin, T., Garipelli, G., Serino, A., Vuadens, P., Tadi, T., Blanke, O., Millán, J.D.: Increasing upper limb training intensity in chronic stroke using embodied virtual reality: a pilot study. J. Neuroeng. Rehabil. **14**(1), 119 (2017)
17. Levin, M.F., Weiss, P.L., Keshner, E.A.: Emergence of virtual reality as a tool for upper limb rehabilitation: incorporation of motor control and motor learning principles. Phys. Ther. **95**(3), 415–425 (2015)
18. Kokkinara, E., Slater, M., López-Moliner, J.: The effects of visuomotor calibration to the perceived space and body through embodiment in immersive virtual reality. ACM Trans. Appl. Percept. **13**(1), 1–22 (2015)
19. Bovet, S., Debarba, H.G., Herbelin, B., Molla, E., Boulic, R.: The critical role of self-contact for embodiment in virtual reality. IEEE Trans. Vis. Comput. Graph. **24**(4), 1428–1436 (2018)
20. Atre, P., Bhagat, S., Pooniwala, N., Shah, P.: Efficient and feasible gesture controlled robotic arm. In: IEEE Second International Conference on Intelligent Computing and Control Systems, pp. 1–6. IEEE Press, Madurai (2018)
21. Badrinath, A.S., Vinay, P.B., Hegde, P.: Computer vision based semi-intuitive robotic arm. In: IEEE 2nd International Conference on Advances in Electrical, Electronics, Information, Communication and Bio-Informatics, pp. 563–567. IEEE Press (2016

22. Kim, H., Tanaka, Y., Kawamura, A., Kawamura, S., Nishioka, Y.: Development of an inflatable robotic arm system controlled by a joystick. In: IEEE International Symposium on Robot and Human Interactive Communication, pp. 664–669. IEEE Press, Kobe (2015)
23. Crainic, M., Preitl, S.: Ergonomic operating mode for a robot arm using a game-pad with two joysticks. In: IEEE 10th Jubilee International Symposium on Applied Computational Intelligence and Informatics, pp. 167–170. IEEE Press, Timisoara (2015)
24. Jiang, H., Wachs, J.P., Pendergast, M., Duerstock, B.S.: 3D joystick for robotic arm control by individuals with high level spinal cord injuries. In: IEEE 13th International Conference on Rehabilitation Robotics, pp. 1–5. IEEE Press, Seattle (2013)

Multithreading Programming
for Feature Extraction in Digital Images

Yair A. Andrade-Ambriz, Sergio Ledesma,
and Dora-Luz Almanza-Ojeda$^{(\boxtimes)}$

División de Ingenierías Campus Irapuato-Salamanca,
Universidad de Guanajuato, Carretera Salamanca-Valle de Santiago km. 3.5+1.8
Comunidad de Palo Blanco, 36885 Salamanca, Guanajuato, Mexico
{ya.andradeambriz,selo,dora.almanza}@ugto.mx

Abstract. Currently, there is a great advance in the construction of processors
with many cores, providing more computational power and resources to use. In
the field of image processing, most of the algorithms use a sequential architecture that prevents from reaching the maximum performance of processors. In
this work, we design and implement a set of low-level algorithms to optimize
the processing of a two-dimensional convolution to obtain the best performance
that a CPU can grant. Our approach uses parallel processing in four different
cases of study based on multithreading. The computation time is compared in
order to find which case achieves the best performance. In the same way, the
computation time of the proposed algorithms is measured, and then, it is
compared with general frameworks, in order to have a real metric of the proposed library with popular Application Programming Interfaces (API's) like
OpenMP.

Keywords: Multithreading · Parallel convolution · Workflow distribution ·
Speedup improvement

1 Introduction

The constant growth of hardware capacity allows the development of image processing
strategies used in important areas as medicine, industry, autonomous systems, among
others. In this context, GPU based approaches provide high speed and take advantage
of similar repetitive routines commonly found in image processing algorithms to
deliver quick results. However, GPU depends on hardware capacity that could be
expensive, and the level of coding is advanced. On the other hand, parallel processing
based on CPU approaches requires only classic programming skills to obtain an
optimal code without having big hardware infrastructure. The CPU-based parallel
processing offers reduced computation time by using the total hardware resources
available.

Nowadays, some programming software like OpenCL [1], OpenMP [2] or
Open MPI [3], among others, provide the framework to program GPU or CPU based
routines using parallel programming Language. However, most of these Libraries or
API's requires complex algorithms and structures and the development time is not

© Springer Nature Switzerland AG 2020
J. Mejia et al. (Eds.): CIMPS 2019, AISC 1071, pp. 208–218, 2020.
https://doi.org/10.1007/978-3-030-33547-2_16

optimal in global. In the case of CPU based routines only, the most common pro-gramming languages are Java, C# and Python, among others, but they also depend on low level commands to support multithreading.

The main idea of this work is to perform image feature extraction in 4 K resolution images for three different kernel sizes using four different parallel programming methods with the main aim to improve the computational performance and reduce the processing time. Each of the proposed methods is based on different a different approach to use multi-thread programming to compute the output of a convolution layer. Finally, our method with the best performance is compared with the OpenMP API and with OpenCL.

This paper is organized as follows. Section 2 presents briefly some related works and describes the feature extraction process based on different ways to implement multithread programming. Section 3 develops the four proposed methods to carry out the multithreading strategy and analyses the performance of each method. Section 4 illustrates the experimental results and the comparative charts for the processing time and the speedup with OpenMP and OpenCL. Conclusions and future works are described in Sect. 5.

2   Parallel Computing Strategy

Digital image processing requires large computational resources from both, software and hardware, to accomplish a specific task. In the case of Convolutional Neural Networks (CNN) a set of images is received as input to perform feature extraction and provide classification results through the convolution processes. GPU and parallel computing solutions have been previously proposed to minimize the processing time during the convolution between images and kernels, and consequently, keep optimal results. However, a GPU is an additional hardware resource that in not always available in some PCs, consequently, CPU parallel computing solutions are more affordable than GPU solutions.

In [4], the authors proposed a CNN implemented on an FPGA (Field Programming Gate Array) which is 31 times faster than the similar approach developed on PC. The process consists of a set of 2D-convolution of parallel primitives using subsampling and no-linear functions for the CNN implementation. The FPGA implementation improves the speedup factor, but the time spent to programming concurrent code is usually high. On the other hand, Cireşan et al. in [5] proposed a CNN train approach for a GPU. Later, Kim et al. 2014 in [6] implemented a convolution process in 2D for a 3x3 kernel using parallel programming demonstrating the high performance provided by multi-threads. However, the increase of logical and physical cores of processors and the increasing emergence of new Application Programming Interfaces (API's) have proposed new strategies to reach lower processing time for the same convolution process. Thus, in [7] Tousimojarad et al. compared the image convolution process implemented in different API's, like OpenMP, OpenCL and GPRM using an Intel Xeon Phi5110P processor.

In this work, we propose a strategy to optimize computational time and extract accurate features in images using three different kernel sizes. Furthermore, the strategy is independent of the operating system and explore four different ways of using the available multi-threads to perform image convolution. Figure 1 shows the global diagram of the proposed approach. First, the process runs the constructor of the unique class used in the code. The constructor sets the values for the initial parameters to three of the main variables proposed: (1) nT processing threads, (2) nC case (3) nK kernels. The nT processing threads variable is set by default to the maximal number of physical or logical cores in the processor. Then, the nC case variable indicates the used method to divide the workflow through the nT processing threads. We establish four different cases (methods) to this variable (they will be described in the next section) by default the variable for the case is set to a method called "one by one". The last variable nK indicates the number of kernels, thus the number of convolutions performed by the process will be nK per image resulting in a nK feature matrix per image. By default, nK is preloaded with an average kernel of size 3×3. Observe that if the user provides his own kernel matrix or other different value for these three variables, the default values will be overwritten.

Once the main variables of the module are defined, the convolution process begins according to the selected case. At the end of the process, the user could choose whether to save the characteristics obtained of the convolution or not. Then, the program releases the used threads, first verifying that they have completed their work and fully stopped, in order to correctly delete their handles and release the used resources, avoiding waste of memory and computational resources. The general diagram has a pre-defined processing block, called "Feature extraction", which runs in accordance with the four cases (or methods) of concurrent processing which are proposed in this study: (1) "one to one", (2) "fixed block", (3) "dynamic block" and (4) "all to one". We want to point out that, the programming strategy used to implement each method was tested for three different kernel sizes and compared among them to find out the method with the best performance, that is, the method with the smallest computational time.

3   Concurrent Convolution Process

In order to obtain the most representative features of an image, we implement a program using a parallel computing strategy that performs a two-dimensional convolution process using different squared kernels of odd sizes from 3×3 to 15×15. Let be A an input image of size $M \times N$ and K any kernel matrix of size $n \times n$, thus B will be the convolved image which is calculated by the Eq. 1 [7].

$$B_{y,x} = \sum_i \sum_j A_{y+i,x+j} K_{i,j} \tag{1}$$

Where x and y represent the position at image A, covered by the K kernel window; i and j move through each location of K and their corresponding value in the image A, the size of B is the same of the input image A. Equation 1 was implemented in a safe-thread function, and therefore, can be called simultaneously from multiple processing threads. Additionally, it can be observed that different strategies can be used to separate the workflow of the program. Being nIm the number of images to be processed, if we have nT processing threads we can divide the convolution task using all the processing threads to process only one image or one thread per image or all the threads to a fixed or dynamic block of data. To carry out the convolution process, we identified four main cases or methods: (1) one to one, (2) fixed block, (3) dynamic block, (4) all to one. Each method performs the convolution process differently depending on the workflow assigned per case.

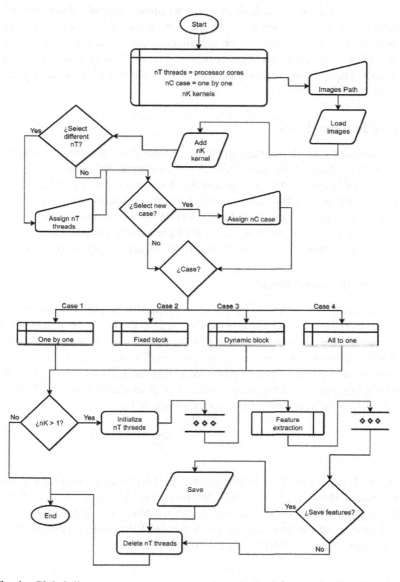

Fig. 1. Global diagram to concurrent extraction of visual features in digital images.

3.1   Case 1: "One to One"

In this method, the workflow is organized so that one thread manipulates one digital image at a time. Once the thread has finished processing the image and the feature matrix is obtained, the next image in memory is assigned automatically to the thread until the total images are completely loaded and processed. When the number of total images processed is nIm, this indicates the end of the task. A global counter $nImG$ is updated each time a thread starts the convolution process, this variable is shared with all the threads, and therefore, it is declared as an *atomic* variable. Each thread uses a local $nImL$ variable to save the index of the current image processed, then the convolution of the total images ends when the $nImG$ is higher than nIm. Furthermore, in the case of nK kernels, each thread processes an image nK times, that is, once for each matrix kernel. The features extracted per image are, then, saved in memory in a vector form.

3.2   Case 2: "Fixed Block"

In this method, the work is separated in blocks of data, then each thread will extract the features to a fixed pre-defined set of images. When threads finish the block of images, the next block of images will be assigned to the threads until the total number of images nIm have been processed. We use two main variables to control which images will be processed per thread and one more fixed variable to control the segment to be processed. As in the previous method, when multiple kernels are provided, each thread will compute nK features images in a fixed block of images assigned to the thread.

3.3   Case 3: "Dynamic Block"

The method called "Dynamic Block" is very similar to the previous method, in this method the workflow is divided into blocks, which they will be processed by nT working threads. The main difference is that the blocks will be assigned dynamically, which leads to a lower collision in memory readings. Theoretically, it allows to achieve a better computational performance when extracting the features. To do this, the total number of images nIm is divided by the number of available threads nT, yielding a balanced division of work among threads as shown in Eq. 2.

$$Workload = \frac{nIm}{nT} \qquad (2)$$

Equation 2 calculates the workload per thread. For instance, suppose that there are 100 images and therefore, nIm is 100. Suppose also that there are four processing threads available, then according to Eq. 2, the workload for each thread will be 25 images. Moreover, if multiple kernels are demanded then each processing thread will extract the features for 25 images multiplied by nK.

3.4   Case 4: "All to One"

In this method, all available threads nT will process one single image at a time. To do this, the height of the image H is divided by nT yielding the size of the section to be processed for each thread; this section is called the *delta* area. Thus, it is necessary to know the value of some variables: the nT or total of processing threads, the current number of the thread cT that is processing the image and the *delta* variable. Figure 2 illustrates how an image is divided during its processing in *delta* areas. Because the size of the block is the same for each thread, you can find the beginning and the end of the block that will be processed in the image.

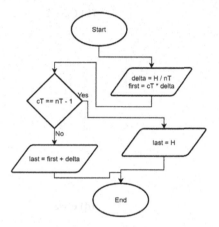

Fig. 2. Obtaining *delta* area on the image.

3.5   Analysis of Performance Per Case

The main benefit of parallel computing is measured by the time it takes to complete a task in a simple process against the time it takes to complete the same task in N parallel processes [8]. The speedup factor $S(N)$ due to the use of N parallel processes is defined in Eq. 3, where $T_p(1)$ is the time it takes for the algorithm finish the task in a single process and $T_p(N)$ is the computation time for the same task using N processes in parallel.

$$S(N) = \frac{T_p(1)}{T_p(N)} \tag{3}$$

For all experimental test on this work we use 100 images with a 4 K resolution and different kernel sizes. Computer simulations were performed using 1 to 8 threads on an Intel i7 7820HQ processor at 2.9 GHz, running Windows 10 pro (1809), DDR4 RAM at 2400 MHz and a NVIDIA Quadro M1200 with CUDA 10.1. Figure 3 illustrates the performance of the four methods proposed in this work with 3 different kernel sizes, it can be observed that the 3×3 kernel has the best performance for all the methods.

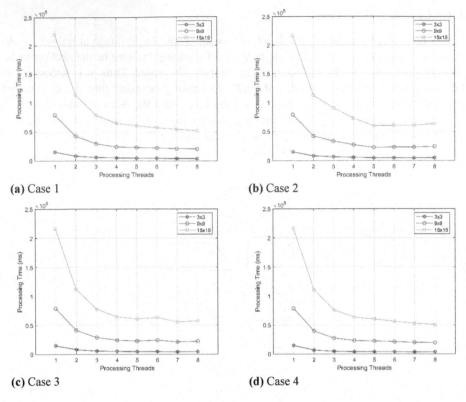

(a) Case 1   **(b)** Case 2

(c) Case 3   **(d)** Case 4

Fig. 3. Performance time of our approach for different kernel sizes per case.

The 3 × 3 kernel size includes a loop unrolling technique to improve the general performance of the methods. Figure 4 shows the processing time and the speedup factor for a 3 × 3 kernel size with a loop unrolling of the total of the kernel size. Also, it can be observed that the method "all to one" achieves the better performance, with the lowest processing time and higher speedup factor.

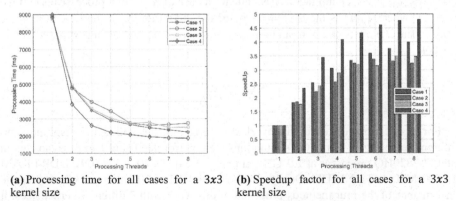

(a) Processing time for all cases for a 3x3 kernel size   **(b)** Speedup factor for all cases for a 3x3 kernel size

Fig. 4. Performance time and speedup for all the methods using a 3 × 3 kernel size.

4   Experimental Results

To optimize the computational time and resources, we designed the class shown in Listing 1 to develop the feature extraction process. This class was implemented using the standard ISO/IEC14882:2011 [9] and ISO/IEC14882:2017 [10], to develop low-level multithreading and for performing operations on file systems respectively. The constructor of the class initializes the private variables, using the maximum number of threads and distributes the work among all working threads.

Listing 1. Main class of the feature extraction using parallel programming.

```
class ConvLayer
{
public:
    ConvLayer();
    ~ConvLayer();
    void loadImages(std::string _path);
    void saveFeatures(std::string _path);
    void addKernel(std::vector<float> _k);
    static void convolution(cv::Mat _img, ... );
    static void convolution(cv::Mat _img, ... );
    void _setCase(size_t _value);
    size_t _getCase();
    void _setThreads(size_t _value);
    size_t _getThreads();
    void process();
private:
    void deleteThreadPool();
    static void C1(std::vector<cv::Mat>& _images,...);
    static void C2(std::vector<cv::Mat>& _images,...);
    static void C3(std::vector<cv::Mat>& _images,...);
    static void C4(std::vector<cv::Mat>& _images,...);
    std::vector<std::string> imagesPaths;
    std::vector<cv::Mat> images;
    std::vector<std::vector<cv::Mat> > convImages;
    std::vector<std::vector<float> > kernels;
    std::vector<std::thread> threads;
    size_t caseP;
    size_t threadsP;
    std::atomic<size_t> nImg;
    std::mutex mx;
};
```

In Fig. 5 it can be seen the extracted features and the original images processed with a 3 × 3 Laplacian kernel using the OpenCV [11] library and our approach. The convolution process converts the data of the input image, to useful information, in this case, it highlights the edges. In addition to feature extraction, a negative processing was performed to improve the visualization of the features extracted on a white sheet.

(a) Original image [11]

(b) Edge feature extracted with OpenCV

(c) Edge feature extracted with our approach

(d) Original image [12]

(e) Edge feature extracted with OpenCV

(f) Edge feature extracted with our approach

Fig. 5. Laplacian filter applied to 4 K image. Image (a) and (d): input image, (b), (c), (e) and (f): edges found using Laplacian kernel, features extracted.

To compare the features obtained with OpenCV and our approach, Table 1 shows the structural similarity (SSIM). The SSIM can take any value from −1 to 1. When both images are identical, the SSIM takes a value of one. We see that the original images have a similar SSIM than those obtained with our approach when compared with OpenCV. Furthermore, the features extracted with our approach have higher SSIM score.

Table 1. SSIM for the features extracted with OpenCV and our approach.

Image		SSIM
Figure 5a vs	Figure 5b	0.29824
	Figure 5c	0.4056
Figure 5b vs	Figure 5c	0.6028
Figure 5d vs	Figure 5e	0.2247
	Figure 5f	0.213
Figure 5e vs	Figure 5f	0.584

Figure 6a shows the comparison for the method of "All to one" described in Subsect. 3.4, with a special case developed in OpenMP, using the same number of processing threads, with a 3 × 3 kernel size, to process 100 images in 4 K resolution.

This special case of OpenMP was designed with the aim of reducing development time using parallel processing. One of the main objectives of the library proposed in this work is to ease the use of multithreading when processing digital images. That is, the user does not need to set any parameter to use multithreading, the user only needs to set the method of workflow distribution. Finally, the results from the computer simulations show that a 20% improvement in the processing time can be observed using the library in this study.

By inspecting Fig. 6b, it can be observed that our method provides a 30% improvement of the performance when compared with OpenMP.

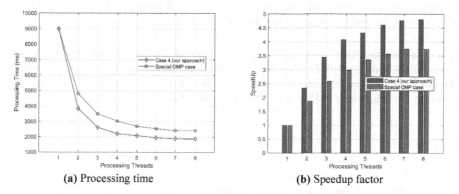

(a) Processing time   (b) Speedup factor

Fig. 6. Processing time and speedup factor for our best case and OpenMP (OMP).

Table 2 summarizes the results for all the methods and devices of the tests of the present work. Additionally, Table 2 shows the processing time in milliseconds to process a batch of images as it was described in Subsect. 3.5. We can see that method 4 "All to one" has the best performance when compared with the methods based on the CPU.

Table 2. Results of processing time comparative of our approach, OpenMP, OpenCL CPU and OpenCL GPU.

Method/Device	One to one	Fixed block	Dynamic block	All to one	OpenMP	OpenCL CPU	OpenCL GPU
Processing time (ms)	2225	2610	2496	1872	2403	1909	1717

5   Conclusions and Future Work

In this work, four different cases to divide the workload between the multiple processing threads of a multi-core processor are proposed. In order to perform a two-dimensional convolution an algorithm was developed and tested. This algorithm carried out the workflow depending on the selected method. Computer simulations show that despite the number of processing threads, the speedup factor improves a little when several threads are simultaneously running to complete one single task. The results showed that the method that offered the best performance was method 4 "All to one", which was compared with OpenMP and OpenCL using the "Loop unrolling" technique to have an idea of the true scope of the work. The speedup factor obtained by our approach provided better results than the special case programmed using OpenMP. The method proposed in this work is easy to use for computing the convolution, and consequently, extract the feature of an image. Furthermore, our method uses only CPU resources. These characteristics are considered the main advantages of our approach. In future work, additional convolutional layers will be included in this same scheme of parallel processing like a CNN structure.

Acknowledgments. This work was partially supported by the project "Fondo Sectorial Conacyt-INEGI No. 290910: *Diseño e implementación de métodos de búsqueda por similitud de usuarios de redes sociales*" and performed during the master degree studies of Yair Andrade funded by the scholarship 634545 granted by CONACYT.

References

1. Open Computing Language. https://www.khronos.org/opencl/
2. Open Multi-Processing. https://www.openmp.org/
3. Open Message Passing Library. https://www.open-mpi.org/
4. Sancaradas, M., Jakkula, V., Cadambi, S., Chakradhar, S., Durdanovic, I., Cosatto, E., Graf, H.P.: A massively parallel coprocessor for convolutional neural networks. In: 20th IEEE International Conference on Application-specific Systems, Architectures and Processors, pp. 53–60, Boston (2009)
5. Cireşan, D.C., Meier, U., Masci, J., Gambardella, J., Schmidhuber, J.: Flexible, high performance convolutional neural networks for image classification. In: Proceedings of the Twenty-Second International Joint Conference on Artificial Intelligence, pp. 1237–1242. AAAI Press, Barcelona (2011)
6. Kim, C.G., Kim, J.G., Hyeon, D.: Optimizing image processing on multi-core CPUs with intel parallel programming technologies. Multimedia Tools Appl. **68**, 237–251 (2014)
7. Tousimojarad, A., Vanderbauwhede, W., Cockshott, W.P.: 2D Image Convolution using Three Parallel Programming Models on the Xeon Phi. CoRR. abs/1711.09791 (2017)
8. Fayez, G.: Algorithms and Parallel Computing. Wiley, New Jersey (2001)
9. I.O. for Standardization: ISO/IEC14882:2011. https://www.iso.org/standard/68564.html
10. I.O. for Standardization. ISO/IEC14882:2017, https://www.iso.org/standard/50372.html
11. Open Source Computer Vision. https://opencv.org/

Selection and Fusion of Color Channels for Ripeness Classification of Cape Gooseberry Fruits

Miguel De-la-Torre[1(⊠)], Himer Avila-George[1], Jimy Oblitas[2], and Wilson Castro[2]

[1] Centro Universitario de Los Valles, Universidad de Guadalajara, Ameca, Jalisco, Mexico
{miguel.dgomora,himer.avila}@academicos.udg.mx
[2] Facultad de Ingeniería, Universidad Privada del Norte, Cajamarca, Peru
{jimy.oblitas,wilson.castro}@upn.edu.pe

Abstract. The use of machine learning techniques to automate the sorting of Cape gooseberry fruits according to their visual ripeness has been reported to provide accurate classification results. Classifiers like artificial neural networks, support vector machines, decision trees, and nearest neighbors are commonly employed to discriminate fruit samples represented in different color spaces (e.g., RGB, HSV, and $L*a*b*$). Although these feature spaces are equivalent up to a transformation, some of them facilitate classification. In a previous work, authors showed that combining the three-color spaces through principal component analysis enhances classification performance at expenses of increased computational complexity. In this paper, two combination and two selection approaches are explored to find the best characteristics among the combination of the different color spaces (9 features in total). Experimental results reveal that selection and combination of color channels allow classifiers to reach similar levels of accuracy, but combination methods require increased computational complexity.

Keywords: Cape gooseberry · Color space selection · Color space combination · Food engineering

1 Introduction

Classification of Cape gooseberry (Physalis peruviana L.) fruits according to their level of ripeness is a repetitive task in the food industry, which requires to provide fruits with high and consistent quality. Current practices to sort Cape gooseberry fruits consist of the visual inspection of color, size, and shape parameters [1]. However, some efforts have been done to improve production methods and provide high-quality products, proposing automated sorting systems based on computer vision techniques. The operation of such systems relies on classification algorithms that consider either different color spaces, or a combination of them. In their most common representation, color is composed of a combination of the three primary colors: red, green, and blue (RGB). The triplet with the values for each primary color is usually considered as a

© Springer Nature Switzerland AG 2020
J. Mejia et al. (Eds.): CIMPS 2019, AISC 1071, pp. 219–233, 2020.
https://doi.org/10.1007/978-3-030-33547-2_17

coordinate system with either Euclidean or a different definition of distance. In such a three-dimensional coordinate system, each point represents a different color. Other three-dimensional color spaces different than RGB are commonly employed, providing different representations that depend on their nature, and can be classified in three categories according to [2]: hardware-orientated spaces, human-orientated spaces, and instrumental spaces. In the first category, hardware-orientated spaces (like RGB, YIQ, and CMYK) are defined based on the properties of the hardware devices used to reproduce. On the other hand, human-orientated spaces (like HSI, HSL, HSV, and HSB) are based on hue and saturation, following the principles of an artist and based on inherent color characteristics. Finally, instrumental spaces (e.g., XYZ, $L*a*b*$, and $L*u*v*$) are those used for color instruments, where the color coordinates of an instrumental space are the same on all output media.

As shown in Sect. 2, the most common color spaces employed in the classification of fruits are RGB, HSV, and $L*a*b*$, which are equivalent up to a transformation. However, it has been shown the accuracy of the same classifier on the same dataset varies from one color space to the other. Some authors have investigated these differences in the classification accuracy due to color spaces or segmentation technique. According to [3] and [4], "In the color measurement of food, the $L*a*b*$ color space is the most commonly used due to the uniform distribution of colors and because it is perceptually uniform." However, it is known that color spaces like RGB, $L*a*b*$ and HSV are equivalent up to a transformation.

Whatever the color space employed in practice, the objective of classifiers applied to fruit sorting consists in finding a criterion to separate samples from one or other class in the so-called feature space. The goal is to establish a decision borderline that may be applied as a general frontier between categories. Supervised learning strategies take advantage of labeled samples to learn a mathematical model that is then used to predict a category in new, never seen, and unlabeled samples. Common supervised classifiers used in the food industry include support vector machines (SVM), k-nearest neighbors (KNN), artificial neural networks (ANN), and decision tree (DT) [5, 6]. In practice, any pattern classifier may be employed, presenting a tradeoff between accuracy and complexity.

Although the equivalence between spaces is well known [7], it has been found that different color spaces provide distinct classification rates [8, 9], and combining spaces through principal component analysis (PCA), an increase in classification accuracy is possible [8]. In a previous publication, authors compared different combinations of four machine learning techniques and three color spaces (RGB, HSV, and $L*a*b*$), in order to evaluate their ability to classify Cape gooseberry fruits [8]. The results showed that the classification of Cape gooseberry fruits by their ripeness level was sensitive to both the color space and the classification technique used. The models based on the $L*a*b*$ color space and the support vector machine (SVM) classifier showed the highest performance regardless of the color space. An improvement was obtained by employing principal component analysis (PCA) for the combination of the three-color spaces at the expense of increased complexity. In this paper, an extension of previous work described in [8] is proposed to explore four methods for feature extraction and selection. The RGB, HSV, and $L*a*b*$ color spaces were concatenated to provide a nine-dimensional color space, from which the most relevant colors are combined and selected by two multivariate techniques and two feature selection methods.

This paper is organized following a sequence that starts with related works and commonly employed techniques, then the exploration of the most popular methods and some experiments to provide evidence of the findings. Section 2 summarizes the most recent works published on ripeness classification, including diverse approaches and methodologies. Some of the most popular methods were selected for this comparison, and Sect. 3 describes the methodology employed for comparison. In Sect. 4, results are described and discussed, and Sect. 5 presents the conclusions and future work.

2   Ripeness Classification

In literature, different color spaces have been employed to design automated fruit classification systems, showing distinct levels of accuracy that are related to both, the color space and the classification algorithm. Table 1 presents common methods and color spaces reported in the literature, employed to classify fruits according to their ripeness level. According to Table 1, the most common color spaces used for classification are the RGB, $L*a*b*$ and HSV. Similarly, the most common classifiers are ANN and SVM. The accuracy obtained by each approach depends on the experimental settings and are not comparable at this point. However, reported accuracy ranges between 73 and 100%.

In Table 1, MDA stands for multiple discriminant analysis, QDA for quadratic discriminant analysis, PLSR for partial least squares regression, RF for random forest, and CNN for convolutional neural network.

2.1   Methods for Color Selection and Extraction

Finding the most relevant features for classification is a crucial step that facilitates the classifier to find the decision borderline between classes. Non-relevant features may conduce to a classification problem that is unmanageable by the most sophisticated classification algorithms. By selecting or extracting a subset of the k most relevant features, redundant and irrelevant data is removed, and a k-dimensional feature space is used for classification instead of the original d-dimensional feature space.

The goal of feature extraction consists of finding information from the feature set to build a new representation of samples. In literature, some feature extraction approaches consider the underlying distribution of samples to find a new feature space that facilitates classification. An example of this unsupervised approaches is PCA, that finds a transformation that maximizes the spread of samples in each new axis. On the opposite, supervised approaches exploit class information to find such a transformation. An example of supervised approaches is the linear discriminant analysis (LDA), that aims to redistribute features and maximize the spread of samples for each class.

On the other hand, the objective of feature selection methods consists of selecting a minimal subset of attributes that facilitates the classification problem. Analogously to feature extraction methods, supervised feature selection strategies consider class labels

Table 1. Color spaces and classification approaches for fruit classification in literature. NA stands for non-available information. The table was taken from [8] and updated with new findings.

Item	Color space	Classification method	Accuracy	Ref
Apple	HSI	SVM	95	[10]
Apple	$L*a*b*$	MDA	100	[11]
Avocado	RGB	K-Means	82.22	[12]
Banana	$L*a*b*$	LDA	98	[13]
Banana	RGB	ANN	96	[14]
Blueberry	RGB	KNN and SK-Means	85–98	[15]
Date	RGB	K-Means	99.6	[16]
Lime	RGB	ANN	100	[17]
Mango	RGB	SVM	96	[2]
Mango	$L*a*b*$	MDA	90	[18]
Mango	$L*a*b*$	LS-SVM	88	[19]
Oil palm	$L*a*b*$	ANN	91.67	[20]
Pepper	HSV	SVM	93.89	[21]
Persimmon	RGB+$L*a*b*$	QDA	90.24	[22]
Tomato	HSV	SVM	90.8	[23]
Tomato	RGB	DT	94.29	[24]
Tomato	RGB	LDA	81	[25]
Tomato	$L*a*b*$	ANN	96	[26]
Watermelon	YCbCr	ANN	86.51	[27]
Soya	HSI	ANN	95.7	[5]
Banana	RGB	Fuzzy logic	NA	[6]
Watermelon	VIS/NIR	ANN	80	[28]
Watermelon	RGB	ANN	73.33	[29]
Tomato	FTIR	SVM	99	[30]
Kiwi	Chemometrics MOS E-nose	PLSR, SVM, RF	99.4	[31]
Coffee	RGB+$L*a*b*$+Luv+YCbCr+HSV	SVM	92	[32]
Banana	RGB	CNN	87	[6]
Cape gooseberry	RGB+HSV+$L*a*b*$	ANN, DT, SVM and KNN	93.02	[8]

to find the most relevant features, whereas unsupervised approaches are based exclusively on the distribution of samples. By nature, selecting the best features subset is a computational expensive combinatorial problem, and their optimality may depend on the strategy followed to rank and/or select the most relevant features. In both cases, feature selection and extraction algorithms, the problem can be stated by considering a set of points (sample tuples or feature vectors) $X = \{x_1, x_2, \ldots x_N\}$, where $x_i \in \Re^d$.

```
Input:
X <- Set of N d-dimensional data samples

    1.  Normalize samples in the d-dimensional dataset
    2.  Compute the dxd covariance matrix C
    3.  Decompose C into its eigenvectors and eigenvalues
    4.  Select the k eigenvectors with k largest eigenvalues (k ≤ d).
    5.  Construct a dxk projection matrix W from the top k eigenvectors
    6.  Transform the samples with X' = X·W
```

Algorithm 1. The procedure followed by PCA.

2.2 Principal Component Analysis (PCA)

PCA is employed to obtain and apply an unsupervised linear transformation that finds the directions of maximum variance in high dimensional data. Such a transformation projects sample patterns onto a new feature space, and the axes with more explained variance typically correspond to those that provide a distribution of samples that facilitate classification.

Algorithm 1 depicts the general procedure to transform data samples from X to their new representation in the k-dimensional feature space X'. The k-dimensional feature space is composed of the k eigenvectors of the covariance matrix C, with highest eigenvalues.

2.3 Linear Discriminant Analysis (LDA)

Like PCA, LDA is employed to obtain and apply a linear transformation that finds the directions of maximum variance in high dimensional data. The main difference is that is supervised and aims to minimize intraclass variability whereas maximizes interclass variability. The number of classes limits the new k-dimensional feature space (*e.g.*, $1 < k < c$, where c is the number of classes). This limitation makes this approach advantageous when the number of classes is elevated, and unpractical for problems with a few classes (*e.g.*, $c \ll d$).

```
Input:
X <- Set of N d-dimensional data samples
L <- Set of labels for the N data samples in X

    1.  Standardize the samples in the d-dimensional dataset
    2.  For each class, compute the d-dimensional mean vector
    3.  Construct the between-class scatter matrix S_B, and the within-
        class scatter matrix S_w
    4.  Compute the eigenvectors and corresponding eigenvalues of the
        matrix S_w^-1 S_B.
    5.  Construct a (dxk)-dimensional transformation matrix W_LDA with the
        k eigenvectors with the k largest eigenvalues as columns.
    6.  Project the samples onto the new feature subspace using W_LDA
```

Algorithm 2. Procedure to transform the sample data into a subspace-based on LDA.

Algorithm 2 presents the procedure to transform the data samples X onto the new k-dimensional feature space using LDA: X'. The k-dimensional feature space is composed of the k eigenvectors with the k highest eigenvalues.

2.4 Eigenvector Centrality Feature Selection

Feature ranking and selection via eigenvector centrality is a graph-based supervised method that ranks features by identifying the most important ones. It maps the selection problem to an affinity graph with features as nodes and assesses the rank features according to the eigenvector centrality (EC) [33].

Algorithm 3 presents the procedure to rank and select the most relevant features from the data samples X. Although this does not constitute a proper transformation in terms of linear algebra, every sample is represented in new k-dimensional feature space with the highest-ranked features.

```
Input:
    X <- Set of N d-dimensional data samples
    L <- Set of labels for the N data samples in X
Output:
    v₀ ranking scores for each of the d features

    1.  Build the weighted graph G = (V,E); where vertices V correspond
        to data samples and edges E among features:
        a.  Compute the Fisher score: f_i = (Σ_{c=1}^{C}(μ_{i,c}−μ_i)²)/(σ_i²), where μ_ij
            represents the mean, and σ_ij is the standard deviation for
            the whole dataset corresponding to feature i.
        b.  Compute mutual information: m_i = Σ_{y∈Y} Σ_{z∈x(i)}  p(z,y)
            log(p(z,y)/(p(z)p(y))), where p(·,·) is the joint probability
            distribution.
        c.  Obtain the adjacency matrix A = αk + (1 − α)Σ, where
            k = f · mᵀ and Σ(i,j) = max(σ^(i),σ^(j)).
    2.  Ranking: Compute the eigenvalues {Λ} and eigenvectors {V} of A,
        where λ₀ = max_{λ∈Λ}(|λ|)
    3.  v₀ is the eigenvector associated to λ₀
```

Algorithm 3. Procedure for supervised feature selection based on eigenvector centrality (ECFS).

```
Inputs:
    X <- Set of N d-dimensional data samples
    K <- Number of clusters, default K=5
Output:
    The top d features according to their MCFS scores

    1.  Construct a p nearest neighbor graph Laplacian (e.g., W_ij = 1 iif
        i and j are connected by an edge; D_ii = ∑_j W_ij; and L = D - W)
    2.  Solve the generalized eigenproblem  Ly = λDy, where Y =
        {y_1,...,y_K} are the top K eigenvectors with respect to the smallest
        eigenvalues.
    3.  Solve the regression problem: min_{a_k}||y_k - X^T a_k||², with a user-defined
        cardinality to control the sparseness of a_k
    4.  Compute: MCFS(j) = max_k |a_{k,j}|
```

Algorithm 4. Procedure for unsupervised feature selection for multi-cluster data (MCFS).

2.5 Multi-Cluster Feature Selection

The unsupervised multi-class feature selection (MCFS) method searches for those features that preserve the multi-cluster underlying structure of training samples [34]. Although the number of clusters (or classes) is unknown *a priori*, it should be explored to find the best feature subspace. The procedure to find the most relevant features using MCFS is depicted in Algorithm 4.

Although the simplest method to choose W was presented in Step 1, other methods exist that range between accuracy and complexity (See [34]). According to the authors, the default number of nearest neighbors is $p = 5$, and the default number of eigen-functions is $K = 5$. This last parameter K usually affects the accuracy of the algorithm and should be optimized before usage.

2.6 Classifiers for Fruit Sorting

According to Table 1, some of the most popular classifiers in fruit sorting are the artificial neural networks (ANN), decision trees (DT), support vector machines (SVM) and k-nearest neighbor (KNN). These classifiers were employed in this paper for the experimental settings. Although these techniques have been present in the literature for a while [35], their usage increased due to their capacity to address real diverse problems.

ANN is a non-linear supervised classifier, that mathematically simulates biological neural networks. A common type of ANN is the probabilistic ANN, which produces an estimated posterior probability for each input sample, to belong to any of the classes. Then, a max function is employed to select the most likely class. In these experimental settings, the Matlab's Neural Network Toolbox was used to implement the ANN classifiers, byways of the *newpnn* function.

DT is a tree-based exemplification of the knowledge used to represent the classi-fication rules. Internal nodes are representations of tests of an attribute; each branch represents the outcome of the test, and leaf nodes represent class labels. In this paper, the Matlab's Machine Learning Toolbox (MLT) was used the train and simulate DTs, using Classification & Regression Trees (CART) algorithm to create decision trees, with the *fitctree* and *predict* functions.

SVM is a supervised non-parametric statistical learning technique that constructs a separating hyperplane (or a set of hyperplanes) in a high-dimensional feature space. Some versions use the so-called *kernel trick* to map data to higher dimensional feature space and find the separating hyperplane there. The functions *fitcecoc* and *predict* functions were used for simulations, both implemented in Matlab's MLT.

KNN is a non-parametric classifier that stores all training samples. The class prediction is based on the number of closest neighbors belonging to a particular class. Given an input sample, the distance to all stored samples should be computed, and then compared, making this classifier complex at the prediction stage. For simulations, the *fitcknn* and *predict* functions from Matlab's MLT were used.

3   Materials and Methods

A set of 925 sample gooseberry fruits with different levels of ripeness was employed in the comparison. Sample fruits were collected from a plantation located in the village of El Faro, Celendin Province, Cajamarca, Peru [UTM: −6.906469, −78.257071]. Fruits were distributed in a 160 cm long, 25 cm wide, and 80 cm high conveyor belt. For image capture, a Halion-HA-411 VGA webcam was employed, with a resolution of 1280×1720 pixels. It was located 35 cm above the sample, and internal walls of the system were covered with a black matte paint to reduce light variations, like [36]. A directional lighting system composed of two long fluorescent tubes (Philips TL-D Super, cold daylight, 80 cm, 36 W) distributed symmetrically on both sides of the sample was used, and a circular fluorescent tube (Philips GX23 PH-T9, cold daylight, 21.6 cm, 22 W) was located at the top. A portable PC was employed for image storage (Intel(R) Pentium(R) Dual-Core CPU T4200 @ 2.00 GHz and 3.0 GB RAM), and a Matlab software was implemented to control the acquisition of the images and their subsequent analysis (Fig. 1).

Fig. 1. Levels of ripeness employed for supervised visual classification, as proposed by Fischer et al. in [37]. Image is taken from [2].

Seven levels of ripeness were employed for visual classification, following the standard for Cape gooseberry, and the visual scale proposed in [37]. The process followed for evaluation is depicted in Algorithm 5.

```
1. Image acquisition, preprocessing, and manual sorting
     a. Image capture of the samples located in the conveyor belt.
     b. Image enhancement with a Gaussian filter.
     c. Fruit segmentation (thresholding).
     d. Manual sorting by color in 6 ripeness levels (by 5 experts).
     e. Organize samples for 5-fold cross-validation.
2. Represent data using the concatenation of the distinct color
   spaces [RGB + HSV + L*a*b*].
3. Extract/select characteristics (PCA, LDA, MCFS and ECFS).
4. Train distinct classifiers with the resulting feature spaces (ANN,
   DT, SVM, and KNN).
5. Test and measure performance.
```

Algorithm 5.   Experimental methodology to evaluate the system with distinct feature extraction/selection methods and different classifiers.

Performance of multiclass classifiers was measured with F-measure, as defined in [8]. First, the confusion matrix is computed according to the responses of each classifier, and true positives (TP_i), false positives (FP_i), true negatives (TN_i), and false negatives (FN_i) are obtained for each class i, using the elements N_{ij} of the confusion matrix. Class-specific precision and recall are computed using Eqs. 1 and 2, respectively.

$$\Pr ecision_i = \frac{TP_i}{TP_i + FP_i} \tag{1}$$

$$Recall_i = \frac{TP_i}{TP_i + FN_i} \tag{2}$$

Finally, the F-measure was used as the reference due to its representativeness of the classification performance on target classes (Eq. 3).

$$F - measure_i = 2 * \frac{Precision_i * Recall_i}{Precision_i + Recall_i} \tag{3}$$

Three distinct analysis were prepared in order to characterize the performance of the system, by fixing the classifier (SVM). First, the best k (number of clusters) were explored in order to obtain the highest classification performance. Then, the size of the feature space was explored in terms of average F-measure. The last analysis explores the performance using the parameters found in previous steps, and the classifiers ANN, SVM, DT, and KNN.

4 Experimental Results

As discussed in Sect. 2.5, MCFS requires a search to find the number of clusters that maximizes the classification performance. The number of characteristics was fixed to seven, to make it comparable with previous results using PCA [8].

Figure 2 presents the box plots that summarize the distribution of performance for the SVM classifier trained with 7 color channels (features) selected with the MCFS algorithm. The parameter that controlled the number of clusters was moved from 1 to 9 (e.g., the maximum number of possible features). In most cases, the median of the F-measure was maintained around 71.75, and only two cases were distinct: 2 and 9. Using 9 clusters seems to provide lower performance related to the generation of an excessive number of clusters. On the other hand, using only 2 clusters for feature selection seems to provide a higher level of accuracy. However, and regardless of the median accuracy, the variability between cases shows a difference that makes no significant difference in using a different number of clusters. In any case, in the experiments described below, the number of clusters is fixed to 2, and other variables are explored.

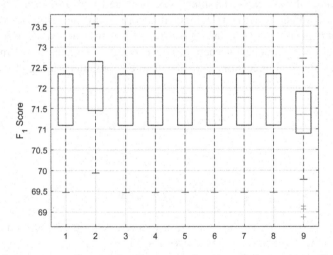

Fig. 2. Boxplots corresponding to the F-measure for nine distinct values of the parameter controlling the number of clusters in MCFS. The number of characteristics was fixed to 7, and the experimentation follows a 5-fold cross-validation strategy.

4.1 Analysis of Feature Spaces

Table 2 presents the average F-measure and standard deviation corresponding to the outputs generated by the SVM classifier after training on feature spaces selected or extracted with the four different methods explained in Sect. 2. d-dimensional feature samples were used for training under a 5-fold cross-validation process.

According to Table 2, experiments showed that using 7 features, either selected with MCFS or combined through PCA, allow for a similar level of performance. ECFS

Table 2. Average F-measure of the SVM classifier applied on distinct feature spaces obtained with the four methods for feature extraction/selection. Bold numbers symbolize the highest F-measure obtained for each method, and numbers in parenthesis symbolize standard deviation.

Method	1-D	2-D	3-D	4-D	5-D	6-D	7-D	8-D	9-D
PCA	40.89	68.56	69.48	71.23	71.83	71.69	**71.99**	71.70	71.65
	(0.34)	(0.91)	(0.95)	(0.82)	(0.92)	(0.70)	**(0.81)**	(0.92)	(0.91)
LDA	52.43	69.10	69.48	70.05	70.02	**71.48**	–	–	–
	(0.81)	(1.24)	(1.17)	(1.00)	(1.05)	**(0.74)**			
MCFS	64.74	65.67	70.04	70.72	71.02	71.92	**71.99**	71.83	71.66
- 2 clusters	(0.70)	(0.68)	(1.13)	(1.04)	(0.96)	(0.76)	**(0.79)**	(0.89)	(0.87)
Color ch.	L*(7)	V(6)	H(4)	b*(9)	R(1)	G(2)	**B(3)**	S(5)	a*(8)
ECFS	40.93	68.81	69.55	71.33	**71.89**	71.76	71.86	71.84	71.66
Color ch.	(0.32)	(1.18)	(1.20)	(0.72)	**(0.72)**	(0.79)	(0.83)	(0.79)	(0.87)
	G(2)	R(1)	a*(9)	b*(8)	**H(4)**	L*(7)	S(5)	V(6)	B(3)

and LDA find their highest performance level with 5 and 6 features, averaging a slightly lower performance than PCA and MCFS. A consistent tendency shows that using more than 3 features allow classifiers to obtain a significantly higher F-measure in all cases, with a lower standard deviation. In other words, in this case using a feature space with more color channels -or features-, either selected (through MCFS or ECFS) or extracted (through PCA or LDA), allows the SVM classifier to obtain a higher and more stable classification performance, at expenses of the evident increase in computational complexity.

In the hypothetic case that only three-color channels were allowed, and these channels could be arbitrarily chosen from the 9 provided by our three-color spaces, in this case, a selection method should be used. Then, a quick look of the 3-D column of Table 2 evidences that in these settings, the MCFS provided a better channel selection, achieving the highest level of F-measure with channels [$L*$, V, H].

4.2 Performance Across Classifiers

The comparative of performance in terms of F-measure, between the ANN, DT, SVM, and KNN classifiers, evaluated on the best d-dimensional feature space found in Sect. 4.2, is presented in Fig. 3. The distinct feature spaces provided a different optimal number of characteristics, and those features were employed in each case. In other words, 7 features were selected for PCA and MCFS, 6 features for LDA and 5 features for ECFS.

As shown in Fig. 3, the SVM classifier overcomes the performance in all cases, and only ANN presents a performance that is close to SVM on the 6-dimensional LDA feature space. In fact, the highest level of F-measure presented by ANN is shown in the space extracted with LDA. In general, in these settings, the performance of all classifiers presents its highest level on the 6-dimensional LDA space. This suggests, that LDA provides a feature space that makes it easier for the work of a classifier after combining information from multiple color spaces. On the other hand, focusing on the two feature selection methods, a similar level of performance is provided by all

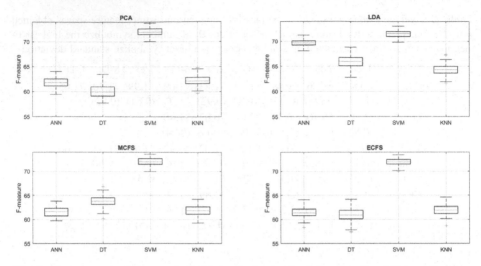

Fig. 3. F-measure for the feature selection/extraction approaches, using four different classifiers.

classifiers, without significant difference. However, if a minimum number of features is required for a given application (*e.g.*, reduce computational complexity and cost), a feature selection method may provide the means to select a few color channels (sensors).

5 Conclusions and Future Work

In this paper, an extension of a methodology was proposed for classification Cape gooseberry according to its ripeness level. As a difference from previous work, two combination approaches (PCA and LDA) and two selection methods (MCFS and ECFS) were explored as feature extraction/selection methods for classification. Test settings include accommodating fruits with distinct levels of ripeness on a conveyor belt and captured with a VGA camera. Segmentation and manual sorting were performed before feature extraction/selection and classification.

Results reveal that selection methods allow classifiers to reach levels of accuracy that are similar, and feature extraction methods require increased computational complexity. The SVM classifier showed superior performance in terms of F-measure regardless of the feature space. In these settings, the classifiers employed in the test presented the highest level of accuracy on the 6-dimensional LDA space.

Future work may include the use of distinct classifiers, and the use of hyperparameter optimization techniques. Similarly, other algorithms for feature selection and extraction may be explored. Finally, another kind of problems may be benefited of selection techniques in food engineering, like using multi or hyperspectral sensors to measure the level of ripeness of Cape gooseberry or any different type of fruit.

Acknowledgments. Authors acknowledge the continuous support provided by the authorities of the *Centro Universitario de los Valles* of the *Universidad de Guadalajara*, as well as the *Universidad Privada del Norte*.

References

1. Zhang, B., Huang, W., Li, J., Zhao, C., Fan, S., Wu, J., Liu, C.: Principles, developments and applications of computer vision for external quality inspection of fruits and vegetables: a review. Food Res. Int. **62**, 326–343 (2014)
2. Nandi, C.S., Tudu, B., Koley, C.: A machine vision-based maturity prediction system for sorting of harvested mangoes. IEEE Trans. Instrum. Meas. **63**(7), 1722–1730 (2014)
3. Du, C.-J., Sun, D.-W.: Multi-classification of pizza using computer vision and support vector machine. J. Food Eng. **86**(2), 234–242 (2008)
4. Taghadomi-Saberi, S., Omid, M., Emam-Djomeh, Z., Faraji-Mahyari, K.: Determination of cherry color parameters during ripening by artificial neural network assisted image processing technique. J. Agr. Sci. Tech. **17**, 589–600 (2015)
5. Abdulhamid, U., Aminu, M., Daniel, S.: Detection of soya beans ripeness using image processing techniques and artificial neural network. Asian J. Phys. Chem. Sci. **5**(2), 1–9 (2018)
6. Hadfi, I.H., Yusoh, Z.I.M.: Banana ripeness detection and servings recommendation system using artificial intelligence techniques. J. Telecommun. Electron. Comput. Eng. (JTEC) **10** (2–8), 83–87 (2018)
7. Schwarz, M., Cowan, W.B., Beatty, J.C.: An experimental comparison of RGB, YIQ, LAB, HSV, and opponent color models. ACM Trans. Graph. **6**(2), 123–158 (1987)
8. Castro, W., Oblitas, J., De-la-Torre, M., Cotrina, C., Bazan, K., Avila-George, H.: Classification of cape gooseberry fruit according to its level of ripeness using machine learning techniques and different color spaces. IEEE Access **7**, 27389–27400 (2019)
9. Bora, D.J., Gupta, A.K., Khan, F.A.: Comparing the performance of L*A*B* and HSV color spaces with respect to color image segmentation. Int. J. Emerg. Technol. Adv. Eng. **5** (2), 192–203 (2015)
10. Xiaobo, Z., Jiewen, Z., Yanxiao, L.: Apple color grading based on organization feature parameters. Pattern Recogn. Lett. **28**(15), 2046–2053 (2007)
11. Cárdenas-Pérez, S., Chanona-Pérez, J., Méndez-Méndez, J.V., Calderón-Domínguez, G., López-Santiago, R., Perea-Flores, M.J., Arzate-Vázquez, I.: Evaluation of the ripening stages of apple (golden delicious) by means of computer vision system. Biosys. Eng. **159**, 46–58 (2017)
12. Roa Guerrero, E., Meneses Benavides, G.: Automated system for classifying Hass avocados based on image processing techniques. In: Colombian Conference on Communications and Computing (2014)
13. Mendoza, F., Aguilera, J.M.: Application of image analysis for classification of ripening bananas. J. Food Sci. **69**(9), E471–E477 (2004)
14. Paulraj, M., Hema, C.R., Sofiah, S., Radzi, M.: Color recognition algorithm using a neural network model in determining the ripeness of a banana. In: Proceedings of the International Conference on Man-Machine Systems, Universiti Malaysia Perlis, pp. 2B71–2B74 (2009)
15. Li, H., Lee, W.S., Wang, K.: Identifying blueberry fruit of different growth stages using natural outdoor color images. Comput. Electron. Agric. **106**, 91–101 (2014)
16. Pourdarbani, R., Ghassemzadeh, H.R., Seyedarabi, H., Nahandi, F.Z., Vahed, M.M.: Study on an automatic sorting system for date fruits. J. Saudi Soc. Agric. Sci. **14**(1), 83–90 (2015)

17. Damiri, D.J., Slamet, C.: Application of image processing and artificial neural networks to identify ripeness and maturity of the lime (citrus medica). Int. J. Basic and Appl. Sci. 1(2), 171–179 (2012)
18. Vélez-Rivera, N., Blasco, J., Chanona-Pérez, J., Calderón-Domínguez, G., de Jesús Perea-Flores, M., Arzate-Vázquez, I., Cubero, S., Farrera-Rebollo, R.: Computer vision system applied to classification of manila mangoes during ripening process. Food Bioprocess Technol. 7(4), 1183–1194 (2014)
19. Zheng, H., Lu, H.: A least-squares support vector machine (LS-SVM) based on fractal analysis and CIELab parameters for the detection of browning degree on mango (Mangifera indica L.). Comput. Electron. Agric. 83, 47–51 (2012)
20. Fadilah, N., Mohamad-Saleh, J., Abdul Halim, Z., Ibrahim, H., Syed Ali, S.S.: Intelligent color vision system for ripeness classification of oil palm fresh fruit bunch. Sensors 12(10), 14179–14195 (2012)
21. Elhariri, E., El-Bendary, N., Hussein, A.M.M., Hassanien, A.E., Badr, A.: Bell pepper ripeness classification based on support vector machine. In: International Conference on Engineering and Technology (2014)
22. Mohammadi, V., Kheiralipour, K., Ghasemi-Varnamkhasti, M.: Detecting maturity of persimmon fruit based on image processing technique. Sci. Hortic. 184, 123–128 (2015)
23. El-Bendary, N., El Hariri, E., Hassanien, A.E., Badr, A.: Using machine learning techniques for evaluating tomato ripeness. Expert Syst. Appl. 42(4), 1892–1905 (2015)
24. Goel, N., Sehgal, P.: Fuzzy classification of pre-harvest tomatoes for ripeness estimation – an approach based on automatic rule learning using decision tree. Appl. Soft Comput. 36, 45–56 (2015)
25. Polder, G., van der Heijden, G.: Measuring ripening of tomatoes using imaging spectrometry. In: Hyperspectral Imaging for Food Quality Analysis and Control, pp. 369–402 (2010)
26. Rafiq, A., Makroo, H.A., Hazarika, M.K.: Artificial neural network-based image analysis for evaluation of quality attributes of agricultural produce. J. Food Process. Preserv. 40(5), 1010–1019 (2016)
27. Shah Rizam, M.S.B., Farah Yasmin, A.R., Ahmad Ihsan, M.Y., Shazana, K.: Non-destructive watermelon ripeness determination using image processing and artificial neural network (ANN). Int. J. Elect. Comput. Syst. Eng. 4(6), 332–336 (2009)
28. Abdullah, N.E., Madzhi, N.K., Mohd Azamudin, A.B.G., Rahim, A.A.A., Rosli, A.D.: ANN diagnostic system for various grades of yellow flesh watermelon based on the visible light and NIR properties. In: 4th International Conference on Electrical, Electronics and System Engineering (ICEESE) (2018)
29. Syazwan, N., Ahmad Syazwan, N., Shah, M.S., Nooritawati, M.T.: Categorization of watermelon maturity level based on rind features. In: Procedia Engineering, vol. 41, pp. 1398–1404 (2012)
30. Skolik, P., Morais, C.L.M., Martin, F.L., McAinsh, M.R.: Determination of developmental and ripening stages of whole tomato fruit using portable infrared spectroscopy and chemometrics. BMC Plant Biol. 19(1), 236 (2019)
31. Du, D., Wang, J., Wang, B., Zhu, L., Hong, X.: Ripeness prediction of postharvest kiwifruit using a MOS E-nose combined with chemometrics. Sensors 19(2), 419 (2019)
32. Ramos, P.J., Avendaño, J., Prieto, F.A.: Measurement of the ripening rate on coffee branches by using 3D images in outdoor environments. Comput. Ind. 99, 83–95 (2018)
33. Roffo, G., Melzi, S.: Ranking to learn: feature ranking and selection via eigenvector centrality. In: Fifth International Workshop on New Frontiers in Mining Complex Patterns. LNCS (2016)

34. Cai, D., Zhang, C., He, X.; Unsupervised feature selection for multi-cluster data. In: Proceedings of the 16th ACM SIGKDD International Conference on Knowledge Discovery and Data Mining (2010)
35. Duda, R.O., Hart, P.E., Stork, D.G.: Pattern Classification. Wiley, New York (2001)
36. Pedreschi, F., León, J., Mery, D., Moyano, P.: Development of a computer vision system to measure the color of potato chips. Food Res. Int. **39**(10), 1092–1098 (2006)
37. Fischer, G., Miranda, D., Piedrahita, W., Romero, J.: Avances en cultivo, poscosecha y exportación de la uchuva (Physalis peruviana L.) en Colombia, Univ. Nacional de Colombia, Technical report, CDD-21 634.7/2005, Bogotá, Colombia (2005)

Model Driven Automatic Code Generation: An Evolutionary Approach to Disruptive Innovation Benefits

Joao Penha-Lopes[1](✉), Manuel Au-Yong-Oliveira[2],
and Ramiro Gonçalves[3]

[1] Quidgest SA, Rua Viriato 7, 1050-233 Lisbon, Portugal
joao.penha-lopes@quidgest.com
[2] GOVCOPP, Department of Economics, Management,
Industrial Engineering and Tourism, University of Aveiro,
Campus Universitário de Santiago, 3810-193 Aveiro, Portugal
mao@ua.pt
[3] INESC TEC, Universidade de Trás-os-Montes e Alto Douto, Quinta de Prados,
5000-801 Vila Real, Portugal
ramiro@utad.pt

Abstract. Of all the technologies born in the 20th century the software industry is the sole one that has kept a large dependency on human physical effort in order to produce the necessary code for an Information Technology system to perform. This paper analyses that evolution in view of today's state-of-the-art regarding automatic code generation based on model definitions for critical business purposes. The social, financial and technological benefits will be addressed as they are disruptive and will affect different levels of society as we know it. Several testimonies were also sought regarding the technology described herein and how it may affect us in our day-to-day lives. We conclude that the software industry is in the midst of a major revolution – which has arrived late in comparison to other industries related to technology and which have decreased the reliance on human labour some time ago.

Keywords: MDD · Model driven development · MDE · Model driven engineering · Automatic code generation · Dependability · Patterns · Meta-modelling

1 Introduction

Since their beginning many industries were heavily dependent on manpower in order to assemble the products they aimed for, the car industry being an excellent example. From the initial hundreds of workers at a plant, to today's robot-equipped assembly lines, there has been a very significant improvement in efficiency. The results and benefits include the decreasing of costs, the decreasing in human error - while increasing throughput. Nevertheless, the software industry has kept a large dependency on human labour. From the tens of people needed to manage the cabling and con-nections of the initial mainframe computers to today's tens or hundreds of

© Springer Nature Switzerland AG 2020
J. Mejia et al. (Eds.): CIMPS 2019, AISC 1071, pp. 234–249, 2020.
https://doi.org/10.1007/978-3-030-33547-2_18

programmers needed, during endless months or years to build complex software solutions, there has been no change in the concept, only in the technology used. This article intends to showcase what is already being done by a 31 year old start-up – Quidgest - which has been implementing disruptive innovation in corporations and governments alike. A true process revolution is already in place and occurring, not tomorrow, but today.

2   A Look at the Literature on the Evolution of Technology

The revolutionary invention of the microprocessor in the late 1960s created a tremendous amount of opportunities in society [12]. New forms of information and communication services appeared and the enthusiasm grew, as concerns the digital world, until the economic downturn registered at the end of the twentieth century [12]. The twenty-first century has not let us down, however, and has brought us surprising change, in terms of how we communicate [13, p. 166]: "Given the growth in usage of online social networks, such as Facebook, YouTube, Instagram and Snapchat, which rely on videos and images (such as photos) to relay information between connections, new intuitive languages, though not yet formally recognized, have emerged". Indeed, social networks hit us hard with Facebook (founded in 2004) and Instagram (founded in 2010), in particular. In turn, the smartphone, brought to the attention of the mass market via the iPhone1, launched by Apple, in 2007, made us more mobile, and we are now accessing and communicating "on the go" i.e. everywhere and all the time. This mobility and digital transformation has been a global phenomenon and has not been limited to the Western world [14].

The unprecedented and profound change, brought about by the Internet, began in the late 1960s as a research project of the US Department of Defence [15]. Since then, firms such as Google now make very significant profits via newly founded industries such as search engines and online advertising. The online giant Amazon has contributed to the proliferation of e-commerce, and a relatively new and growing concern has been that of making this distribution channel accessible to all, irrespective of our differences [16–18]. Education has also been changed by the new digital reality, as the younger generation has suffered a deep social impact brought on by technology, as the newcomers to firms, the so-called millennials, want to be online 24-7. "Millennials interact with technology like no other generation before them and this is affecting how they want to be taught in higher education and how they want to lead and expect to be led in organizations, after graduating" [19, p. 954]. It is in this context that the firm Quidgest has appeared, standing for a profound change in how we program our computers, how we communicate our "commands" to them, one of the major and most important tasks in society today. Just as we saw with operating systems and with MS-DOS (dominated by lines of code) and its transition to the Windows environment, so we predict shall occur with mainstream computer programming and its increased user-friendliness.

3 Methodology

The authors looked into the main players of software development, both in the areas of product development, such as Windows- and Linux-based frameworks as well as the main consumers of such platforms such as the developed countries, known as belonging to the digital group D9. The corporations that provide software solutions were also analyzed, not only those following the traditional pattern of programming, but also those following the recent technology of low-code. The authors also found testimonies of individuals who have been confronted with the software development technology Genio, described in this paper.

4 Problem

The problems identified today concerning traditional software solution development are several and each of them will be addressed individually. The purpose is to highlight the hidden damage that keeping this methodology will bring to organizations and countries. If a pain is there for too long no one will notice it anymore.

4.1 Governments Are Inducing Entropy

As governments claim their countries are going digital, there is now a group of Digital Nations known as D9 [1], which include Canada, Estonia, Israel, Mexico, New Zealand, Portugal, South Korea, the United Kingdom and Uruguay.

The purpose of this group of digital countries is to spread world-class digital practices, to identify and improve common problems and, as an ultimate objective, to grow digital economies. In order to achieve these purposes the D9 group of countries signed, on the 22nd November 2018, in Jerusalem, a charter of nine principles:

1. User needs: the design of public services for the citizen
2. Open Standards: a commitment to credible royalty-free open standards to promote interoperability
3. Open source: future government systems, tradecraft, standards and manuals are created as open source and are shareable between members
4. Open markets: in government procurement, create true competition for companies regardless of size. Encourage and support a start-up culture and promote growth through open markets
5. Open government (transparency): be a member of the Open Government Partnership and use open licenses to produce and consume open data
6. Connectivity: enable an online population through comprehensive and high-quality digital infrastructure
7. Teach children to code: commitment to offer children the opportunity to learn to code and build the next generation of skills
8. Assisted digital: a commitment to support all its citizens to access digital services
9. Commitment to share and learn: all members commit to work together to help solve each other's issues wherever they can

Point 7 of this list of principles passes on the need to teach children to code and this is precisely the added entropy to the system; it represents the perpetuation of the need for human labour to execute tasks that can be fully automated with the added disadvantage that the coding technology is ever changing, meaning that the children, during their childhood will have, most probably, to learn more than one programming language and this is a waste of time and energy.

Global consulting experts, such as Gartner [2], state that by 2022 40% of all new application development projects will be executed with the help of artificial intelligence (AI).

4.2 Today's Business Models Present Resistance to Change

Software solutions have always been developed by companies hiring a large number of programmers, which has proven to be a very profitable business model, with large consultancy firms earning revenue by the hour (of human labour), often over years at a time, to produce customized software suites which they then made inaccessible to other software firms. On the other hand, these programmers type in the code lines necessary to make the solution work and although nowadays programming is structured the fact is that there is always resistance from programmers to analyse and modify code created by someone else. These companies have their business models built around this reality: there are consultants that collect the customer's business rules required for the application (the so-called strategists) and, then, there are the programmers that create the necessary code to comply with those rules.

There are also companies that, not developing any code, rent to their customers the necessary programmers in order for them to develop whatever the customer needs. In this case the business model is selling time, the time multiplied by the number of programmers allocated to the customers.

Both of these contexts are well established and there is a large resistance to change by both of them. This resistance associated with the shortage of software developers which is estimated to be as large as 1M for 2020 [3] will allow to justify the maintaining of developing costs at a very high level, which is always a good motivation for keeping business as is.

4.3 Cognitive Ignorance

Usually when people face new concepts they tend to create referrals to concepts already known and understood, but when this new concept is unique, all efforts for a "similar" concept are doomed and either the person feels lost or is attached to the wrong concept. Considering the technology described in this paper, it is often associated with "low-code" when, in reality it has nothing in common (Table 1).

Table 1. Avenues of software development code in the creation of problems.

Avenue	Comment
Time	Time spent by programmers
Error	Errors introduced by humans
Changes	Black-box not flexible
Cost	High cost due to time and expertise
Upgrade from legacy	A pure manual and utterly difficult approach
Complexity	Exponentially increasing
Inclusion	None

4.4   Time

Time is, indeed, a problem mainly for large groups implementing large projects such as an ERP. An ERP may have to manage a number of different companies in areas such as Finance, HR and Asset management. Furthermore, these companies may want to keep their individual management, although the group needs management Key Performance Indicators (KPIs), in real time from the duly processed information available at each company. It is expected that these implementations may take years and sometimes cost hundreds of millions of Dollars. It is thus expected but it does not have to be that way, as we will see later on in this paper.

4.5   Error

Humans are prone to err, it is in our nature. The reasons for errors in the coding phase may be due to a number of reasons where the most common are:

- poor understanding of the customer's need by the consultant that created the requisites for the solution;
- poor architecture of the software solution made by the programmer;
- programming errors. Any error is very difficult to track and correct, let alone the fact that there is this consultancy structure in between the customer and the programmer.

4.6   Changes

Today's dynamics of global business would be unthinkable just half-a-dozen years ago. Not only do business practices change regularly at ever decreasing cycle times due to global competition, but also legislation in different countries, in different areas of the world keeps changing in order to adapt to the always faster technological advances we are witnessing. The banking industry has been one of the areas where legislators recently introduced more changes namely around regulatory reports.

Out-of-the box products, no matter how many variables they have to be parametrized, are fixed outside those capabilities. Different business rules will imply deep changes in the code; if these changes are mandatory by legislation the vendor will have to execute them; but if they are for the convenience of the customer, then they either will cost a great deal or will not be executed at all.

4.7 Cost

Cost is the bottom line factor but, interestingly enough, people and organizations do have the expectation to pay high values for complex and sophisticated software solutions. As we have seen above there are arguments to support that thesis, such as:

- programmers are expensive
- due to the fact that there is a shortage of them
- the customer asked for a complex solution
- which implies many programmers for a long period of time.

4.8 Upgrade from Legacy Solutions

The larger the organizations the more probable it is to find legacy solutions, i.e., software applications that are already obsolete but, as they are working, everyone is very scared of changing them due to the risks it implies. This applies to large organizations or governments. There are different difficulties when upgrading from a legacy solution:

- managing the change to the new application for all the users that were used to the old one
- migrating the data from the old one to the new one and
- creating the code for the new solution to allow for the functionality the customer wants to maintain as well as all the new features that are necessary to introduce with the new technology.

4.9 Complexity

The increasing complexity of organizational ecosystems, progressively built around web-based supported solutions and architectures, places growing demands on the development of omnichannel software along multiple perspectives: functionality, interoperability, security, availability, ubiquity, quick response to change and evolution, among others. Furthermore, all these perspectives are interconnected at the time of creating the code and most of the times the code created to satisfy one may contradict another and, thus, code must be changed, documented and none of it is done quickly or cheaply.

4.10 Inclusion

Inclusion is the great miss from the programming activity. More and more the technologies grow in sophistication; more and more the complexity of the requirements to process critical business data grow exponentially and, hence, using these ever more complex technologies for these ever more sophisticated environments is an absolute exclusion factor for the large majority of the people who do not have a technological background.

5 Solution

So far we have approached the issue of traditional programming by creating code while punching keys at a keyboard. This is an activity that not only demands hard human labour, but also requires an ever changing knowledge of technology. The associated problems have been summarized in the previous section. Attempted solutions have been showing up in the market, in recent years, but either due to their business or technological models the end result has not reached the level of the solution presented here today. Reasons for this distance have to do with the fact that the created code, of those attempted solutions, has to reside in the platform where it has been created, for a lifelong time fee, and those platforms do not allow for large and complex software solutions and with a large number of them, known as low-code a great deal of technological background is still needed.

5.1 Need Drives Great Innovation

Since the creation of Quidgest, back in 1988, the founders have been challenged in relation to time management and the efficiency of human resources both internally when creating solutions, as well as externally to satisfy the needs of the customers. One of the first large projects was related to agro-investment analysis, a very lengthy process of data analysis which only allowed each expert to analyse 0.6 projects each day.

While developing the software by the traditional programming way of learning a language and typing in the code, it was soon realized that there were a number of patterns repeating themselves in the code; all sorts of patterns which we will detail later on, and these patterns were taking a tremendous amount of time to keep repeating the very same code every time the same pattern needed to be inserted; this was a huge waste of efficiency.

It was at this point that the decision to build a platform that could automatically generate the code for those patterns was taken and development started. From that moment onwards all the solutions were built using this platform, designated as Genio, an acronym to GENerate Input Output, and that Genio has grown to 31 years old with hundreds of patterns, already embedded.

The agro-investment referred to above was finalized in record time and allowing each expert to go from 0.6 projects/day to 25 projects/day. The record time when building solutions, with the support of Genio has been kept and it is, still today, 1/10 (one tenth) of the time normally accepted by customers and using only 1/10 (one tenth) of the usual necessary resources. Ingenuity allows Genio to perform 100 times better than traditional programming and without any of the problems associated with it.

5.2 Patterns at the Core

It is interesting to consider that although Genio is already 31 years old, today's researchers such as Hamid and Perez [4] propose a pattern-based development approach to address dependability through a model-driven engineering approach, precisely what Genio has been configuring since its creation.

As stated, patterns are at the very core of the automatic code generating platform Genio and these patterns come in all shapes and sizes:

1. Persistence and Data Structure patterns
2. Patterns related to Processes and Workflow
3. Business Logic patterns
4. Patterns related to Interfaces and User Experience (UX)
5. Security and Audit patterns
6. Debugging, Testability and Robustness patterns
7. Patterns for System Administration
8. Patterns for the International and Global use of systems
9. Patterns that ensure Integration and Interoperability
10. Patterns that support Software Engineering and Lean Management (SE4.0)
11. Specific patterns for a function or business area (inc. Machine Learning)
12. Patterns for Extensibility
13. Patterns simplifying the Transition to Genio Models

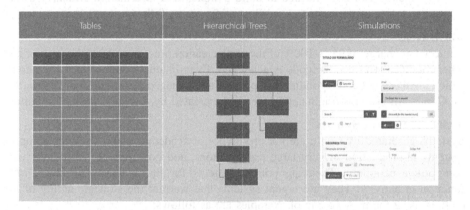

Fig. 1. Different means to represent the different types of patterns at Genio Model IDE.

Internally, however, and regardless of the different IDE representations (Fig. 1), all the Genio Model components persist as tables. Therefore, Genio itself is not really different from solutions made by Genio. In fact, Genio is modelled and generated by Genio. That means Genio is a second generation generator. The metamodel or metadata model used by Genio to build even the most complex integrated systems is relatively simple and easy to understand. The Genio Model is also

1. an evolutionary "language" (it is possible to add new attributes/patterns to the model)
2. a multi-tenant model (through characteristics defined for each customer, different solutions are generated, although common procedures are shared)

Finally, each Genio pattern includes the corresponding generation process, through inference, translating the model into the solution source code.

5.3   Models at Work

Genio adopts a MDE[1] approach, from which a technologically independent information architecture emerges, but keeping a close interaction with that technology. The drawing of the models does not require any technological background, since only a good knowledge of the business is necessary. We do have four types of models that are created in Genio to cover all the functional and informational needs:

- Data model
- Interface model
- Workflow model
- Business model.

5.4   Principles, Characteristics and Guidelines

In order to have these models defining the behaviour of Genio there are a number of principles, characteristics and guidelines that have been implemented to automatically generate the lines of code needed to build complex and information critical software solutions. A few of them, just as an illustration, are as follows:

100% Genio
Genio takes charge of 95%–100% of the code creation process without the need of any other modification once the code is generated. Even manual components of the code, which are those not corresponding to a pattern, are managed, maintained and controlled by Genio. All systems originated in Genio are created by, literally, pushing a button.

Real Life Critical Systems
Systems generated by Genio are not prototypes or academic trials. They are complex systems in production in very large organizations, and governments, in today's very competitive economy. They are critical systems for the specific procedures of a function or activity. They are urgent systems that warrant the strategic agility of the entities using them. A couple of examples are as follows:

1. Nova University (Lisbon, Portugal) (UNL): Nine faculties, each with its own ERP + an overall foundation that needed to have integrated business data from all of them. The solution provided allowed for only one ERP for the whole university while maintaining the operation's individuality of each of the faculties. UNL publicly praised [5] the fact that only three months after the order parts of the new ERP were already in production.
2. Full HR management for the Jamaican Government: 120 000 staff distributed over 200 entities. The whole project was created by only 4 people during 6 months. The Jamaican Government published a video motivating their staff to use the solution [6].

---

[1] Model Driven Engineering.

Simulation

Since all models are a simplified representation of reality, Genio uses simulation as the closest approach to reality. Simulation ensures that the gathered metadata is the one necessary for the creation of each system. The functional requirements are listed defining, directly in Genio, the necessary characteristics of the simulation and, at the end, the final system.

Hyper-Agile

Very fast prototyping, interactivity, very short cycles (hours, not months), the development of practices with the customer organization, the promotion of the integration of the customer's perspectives (crowdsourcing) and a quick response to change, do assure the compliance of Genio with the Manifesto for Agile Software Development [7].

Independent Layers for Technology and Functional Specifications

The modelling of the system, during which the functional requisites are listed, is a distinctive phase from the pattern creation that, in turn, will support the code generation. Thus, each layer, technological (patterns) and functional (models) evolves autonomously, while managed by different teams and addressing different needs.

1-click (Re)generation

Any Genio user, even if not familiar with the project, can now, or at any moment in the future, re-generate the whole code by pressing a button. To achieve this it is only necessary to access the models drawn for the project, with any Genio version. This is an advantage that assures the best agility and maintenance simplicity. Under this principle one does benefit from a significant lowering of the effort required by any business rules change, with an increase in the system's stability, as well as the continuous integration of new technologies as they appear.

Genio's training reflects precisely this concept of the no need for any kind of programming knowledge. The formal training may engage different levels [8], but for the sake of argument we will only analyse Level 1.

Goals:

1. Become aware of the model driven approach
2. Generation requirements in Genio
3. Know the core patterns in Genio
4. Generate fully functional applications with Genio

Pre-requisites:

1. Web browser use
2. General Windows user knowledge

Training Program:

1. Initiation
   a. Base knowledge and vision
   b. Explain what Genio is
   c. Install and configure Genio

2. Core
   a. Draw models for a Relational Database (tables, domain, areas, relations)
   b. Draw models for the Interface (forms, menus and lists)
3. Genio project
   a. Identify requirements
   b. Apply knowledge
   c. Present Results

Certification:
The certification requires a minimum score of 70% as an average of continuous evaluation, project presentation and final exam.

Error Diminishment
The re-use of the same code patterns, in several instances in a project, by different users and at different customers as time flows will diminish significantly the probability of errors in the code generated by Genio. This cycle is fully managed, measured and optimized. Any error detected in the code generated by Genio is reported to R&D and immediately taken care of by correcting the patterns or the generating routines.

Supported Technologies
The models can be instantiated in several technologies, usually those that at a point in time are the most used and convenient. Today they are:

- Microsoft .NET MVC
- Microsoft C#
- Microsoft MFC
- Office Add-in
- Android
- HTML 5
- MySQL
- Oracle
- Windows Azure
- Microsoft SQL Server
- Java

Customer software solutions must be kept using the latest version of whatever technology was used to generate the code. Genio can easily support the motto "Microsoft launches in the morning and Genio integrates in the afternoon". Genio will support the technology as long as it is technically possible.

Model Coherency
It is required of the Knowledge Engineer that he knows and is able to test the coherency of the necessary changes to an existing project, based on the core models used to create that project. In traditional programming the extreme difficulty to introduce changes works as pure resistance to change[2].

---

[2] A change in the data model, for instance, will imply a need to review every line of code written.

5.5   Function Points

The function points express the quantity of information processing that an application provides to a user [9]. The ISO/IEC 24570:2018 sets the standards and conditions for this evaluation. Put to the test we have compared Genio with the productivity of both manual programming and low-code tools (Fig. 2).

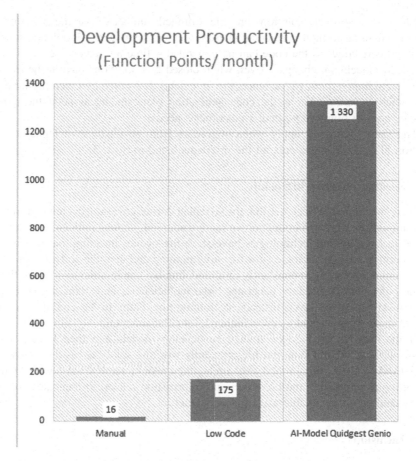

Fig. 2. Monthly delivered function points depending on the code generation tool.

6   Benefits

The benefits that derive from the advantages allowed by Genio are not only related to software development but also, and very much so, to Corporate, Government and Society disruptive evolutions. Whatever the type of benefits shown, they will all be converging to a better quality of life, individual or corporate-wise.

6.1   Solution's Flexibility

Organizations develop their best practices along many years of experience, no matter if they are corporations, governments or any other sort of organization. Time allows for the refining of methodologies and procedures and, therefore, it is not efficient to use an application just because it claims to have "embedded" the best practices for a particular business. There is no such thing as a one-size-fits-all in what regards operational best practices [10].

Closed-box software solutions provide a limited number of parameters to be configured according to their core code whereas Genio's generated solutions do not have any fixed-core-code, as the code can be generated – time and again – every time the associated models are changed. Even when closed-box solutions have to be modified due to legal imperatives, for instance, the time delay and costs are much higher than if those changes had Genio as the code generating platform, let alone if the required changes are just due to a particular customer's needs.

The same arguments apply when upgrading from an obsolete or legacy existing solution [11], since it prevents all the problems listed in Sect. 3.

6.2   Seeding Future Citizenship

We have mentioned in Sect. 4.1 that the so called digital governments are committed to teaching children how to code. As we have discussed, coding implies learning some sort of technology and technologies have short life cycles, meaning that one needs to keep learning new technologies just for programming and then the software, created in the meantime, needs to be constantly modified, line-by-line in order to adapt to the new technologies. This procedure generates large inefficiencies and trains the brain to do mechanical repetitive tasks instead of training the brain to be creative in model development which lies at a much higher level of abstraction.

If the children were to be trained to develop models and then have a tool to transform those models into real life efficiently working solutions, as they grow to be specialized in their own professional areas, they would always have with them the capability of automating the processes and information related to their areas of work. This would be an invaluable asset for any society.

6.3   Inclusion

Most people do not have the technological background to sit and manually use a computer to create a program for a specific solution that they thought about. Having stated this it does not mean that those very people are not able to precisely define what they wanted the solution to do, based on what inputs and having the processing made in a certain way; in short, they cannot program but they do know the business rules.

If these people were given the possibility of creating their own solutions the efficiency improvements within large organizations and governments would be exponential and clearly disruptive, as it would be if the unemployed people with experience in some sort of activity were to be empowered with the capability of automating the knowledge they have. Imagination is, literally, the limit.

7 Testimonies

In the following paragraphs three anonymous testimonies are provided[3], concerning the disruptive technology addressed in this article.

7.1 Father of a 16 Year Old Boy (Lisbon, 2019)

"During the summer of 2018 my son went to do Genio's first level of training. During one week - mornings and afternoons. Because the training was given free of charge my son compensated by doing a couple of weeks of work at Quidgest in the area of document and GDPR management. He was assigned the development of a module for an existing customer, a private bank. Three weeks later I had a very proud boy arriving home and stating: "Daddy, today I went [with his Quidgest supervisor] to the bank and successfully showed the new module working; in a meeting where the bank's CEO was present!""

7.2 C-Level Executive (Porto, 2019)

"Programming for me has always been a problem. My father, back in 1986, predicted that computers would be the future. As a consequence, I decided to do a degree in that area. Albeit, the programming degree I enrolled in did prove challenging in some aspects. Although I did the more theoretical subjects with relative ease, the programming courses proved to be a problem. My programs never seemed to work, though I had a good grasp of the theoretical concepts. I had some help and tuition from older colleagues, however, I never fully mastered the programming technique, in whichever programming language I tried. After graduating I was offered jobs as a computer programmer, as I was still a top student from my degree. I did not have, however, the self-confidence in my abilities as a programmer to take on computer programming jobs. This was even true after I was admitted to Andersen Consulting, in the late 1990s, as a consultant. I stayed away from more technical tasks, despite Andersen Consulting, at the time, being an accomplished customised software producer. When I recently heard about Quidgest and about their radically innovative invention of a process to forego much of the technical programming aspects, I was delighted. Finally, I saw, I could fulfil a lifelong dream of making a computer do as I wish, via the automatic generation of software".

7.3 MSc Student in Engineering and Information Management Systems (Braga, 2018) [8]

"It was necessary for me to develop a platform for the support of an R&D unit at my university and, thus, the possibility of being trained in Genio just popped-up. This training was essential to allow for the understanding of the development methodology as well as providing for the exploring of the deeper functionalities of Genio. Since this

---

[3] If need be, the persons to whom they relate may be available to cross-check their authenticity.

was my first experience with a code development tool, based on high level specifications, Genio's potential and its speed to create solutions left me positively baffled. The training was very useful and productive for the four days it lasted. Actually, before the training I had no idea that such significant progress could be made concerning application development, in such a short time. I really appreciate the excellent professionalism of the trainer".

8 Conclusion

Software development has been kept within the same human dependent operational concepts for decades now, due to the installed business models and economic interests and the general ignorance of the state-of-the-art, as far as customers go. This state of affairs implies huge costs to customers, not only in the direct value of pricing but also the costs associated with very long implementation times.

Today, not in the future and not only in some testing scenarios, it is already possible to create large and complex business-oriented solutions solely based on model drawing and letting a platform generate the hundreds of thousands or millions of lines of code which can be transported anywhere in the world in order to be executed.

This is the kind of advanced technology that surely benefits Humanity.

References

1. New Zealand Government. https://www.digital.govt.nz/digital-government/international-partnerships/the-digital-9/. Accessed 15 May 2019
2. Prepare for Automation's Impact on Application Development. https://www.gartner.com/smarterwithgartner/prepare-for-automations-impact-on-application-development/ Accessed 20 May 2019
3. The Talent Shortage of Software Developers in 2019. https://fullscale.io/the-talent-shortage-of-software-developers-in-2019/. Accessed 20 May 2019
4. Hamid, B., Perez, J.: Supporting pattern-based dependability engineering via model-driven development. J. Syst. Softw. **122**, 239–273 (2016)
5. Um grande projeto na NOVA: Quidgest. https://www.linkedin.com/posts/ana-rita-marante-0796a729_caso-de-estudo-erp-singap-universidade-activity-6487620546635792384-F7BP. Accessed 30 July 2019
6. MyHR+: Connectivity in the public sector is happening now. It's here to stay! https://www.youtube.com/watch?v=oV1HQkGEykA&list=PL2zql1VAlN-3St3i-IFP23t_qngds36-b&index=2&t=3s. Accessed 30 July 2019
7. Manifesto for Agile Software Development. https://www.agilealliance.org/agile101/the-agile-manifesto/ Accessed 28 June 2019
8. Quidgest Academy. https://quidgest.com/en/quidgest_academy/. Accessed 30 July 2019
9. Definitions and counting guidelines for the application of function point analysis. https://nesma.org/wp-content/uploads/edd/2018/07/EN-viewer.pdf. Accessed 20 May 2019
10. Major Myths Dragging Down Your Software. https://documentmedia.com/article-2877-6-Major-Myths-Dragging-Down-Your-Software.html. Accessed 28 May 2019
11. Upgrading from [Obsolete] Solutions. https://www.linkedin.com/pulse/upgrading-from-obsolete-solutions-joao-penha-lopes/. Accessed 28 May 2019

12. Mansell, R., Avgerou, C., Quah, D., Silverstone, R.: The challenges of ICTs. In: Mansell, R., Avgerou, C., Quah, D., Silverstone, R. (eds.) The Oxford Handbook of Information and Communication Technologies, pp. 1–28. Oxford University Press, Oxford (2009)
13. Au-Yong-Oliveira, M., Moutinho, R., Ferreira, J.J.P., Ramos, A.L.: Present and future languages – how innovation has changed us. J. Technol. Manage. Innov. **10**(2), 166–182 (2015)
14. Au-Yong-Oliveira, M., Branco, F., Costa, C.: A evolução cultural graças à adoção da tecnologia e ao fenómeno "Mix-Tech" – Um estudo exploratório baseado na observação. RISTI – Rev. Ibérica de Sistemas Tecnol. Inform. **17**, 854–869 (2019)
15. Greenstein, S., Prince, J.: Internet diffusion and the geography of the digital divide in the United States. In: Mansell, R., Avgerou, C., Quah, D., Silverstone, R. (eds.) The Oxford Handbook of Information and Communication Technologies, pp. 168–195. Oxford University Press, Oxford (2009)
16. Gonçalves, R., Martins, J., Pereira, J., Oliveira, M.A., Ferreira, J.J.P.: Enterprise web accessibility levels amongst the Forbes 250: where art thou o virtuous leader? J. Bus. Ethics **113**(2), 363–375 (2013)
17. Gonçalves, R., Rocha, T., Martins, J., Branco, F., Au-Yong Oliveira, M.: Evaluation of e-commerce websites accessibility and usability: an e-commerce platform analysis with the inclusion of blind users. Univ. Access Inf. Soc. **17**(3), 567–583 (2018)
18. Silva, J.S., Gonçalves, R., Branco, F., Pereira, A., Au-Yong-Oliveira, M., Martins, J.: Accessible software development: a conceptual model proposal. Univ. Access Inform. Soc. (UAIS) **18**, 703–716 (2019)
19. Au-Yong-Oliveira, M., Gonçalves, R., Martins, J., Branco, F.: The social impact of technology on millennials and consequences for higher education and leadership. Telemat. Inform. **35**(4), 954–963 (2018)

Information and Communication
Technologies

Cluster Monitoring and Integration
in Technology Company

Veronica Castro Ayarza[2] and Sussy Bayona-Oré[1,2(✉)]

[1] Universidad Autónoma del Perú, Lima, Peru
sbayonao@hotmail.com
[2] Universidad San Martin de Porres, Lima, Peru
vcastroayarza@gmail.com

Abstract. Companies currently use cluster technologies as a platform for applications that support their core business services. Implementing clustering technologies ensure high availability for mission-critical applications and services, because both hardware and software failures are quickly detected. The operations of a failed node are immediately resumed by another node in the cluster Hidden risks or failures are inevitable in any IT environment. No enterprise is exempted from its clusters being able to suffer inconveniences that will generate unavailability of service and in turn financial losses. Monitoring the infrastructure allows better use of resources, proper operation, and system performance. This article describes the implementation of a monitoring system named Zabbix and applying a network design methodology called PPDIOO. Implementation will reinforce the monitoring of clusters and customizes the monitoring system. It will increase customer satisfaction, rules for cluster events and monitored resources.

Keywords: Cluster · Zabbix · Monitoring · PPDIOO

1 Introduction

Today information and communications technologies (ICT) plays a strategic role in the modern enterprise [1] because it supports different activities and services offered. These services are supported by applications, database-network architecture, and storages (server platforms). The network architecture of a modern company has four basic characteristics which are: fault tolerance, scalability, quality of service and security [2]. For most critical services a platform of a set of servers known as clusters is used, which ensures high availability of services offered. This availability in some cases may be absent, with downtime, because hardware or software failures that are inevitable in any IT environment [3].

A cluster is a group of independent computers that work together to increase the availability and scalability of services and applications. If one or more cluster nodes fail, other nodes begin to provide service. That is, if a server or cluster node were to fail besides causing a lack of availability of service and financial losses for the company, it triggers a series of much larger problems if more servers depend on it. To the extent of leaving unworked websites, applications or incomplete systems. Just to be out of

© Springer Nature Switzerland AG 2020
J. Mejia et al. (Eds.): CIMPS 2019, AISC 1071, pp. 253–265, 2020.
https://doi.org/10.1007/978-3-030-33547-2_19

service, for at least five minutes can not only cause financial losses but loss of reputation for the firm [4]. Failure rates also begin to increase significantly as servers age with time. Two-year servers have a 6% failure rate while seven-year servers have an annual failure rate of 18% [5]. Failures can be prevented through proactive monitoring by having alerts or reports [6, 7] when degradation is about to occur and allows to perform corrective actions. It is for this reason that organizations need monitoring of the infrastructure systems that facilitate administration and detection of anomalies [8]. In this way it is possible to prevent incidents and detect problems as soon as possible, as well as saving costs and time.

This article describes the experience of implementing the monitoring system in a technological company to establish corrective measures before systems are disrupted. In this case, after evaluating the situation of the company, it was found that the current system of monitoring infrastructure does not monitor all the resources that make up the cluster. In this case, after assessing the situation of the company, it was found that the organization had shortcomings in the monitoring infrastructure system used, because they did not provide all the resources that make up the cluster and a detailed status for each. These weaknesses in monitoring resulted with no certainty of which clusters were not properly configured. Why not proper operation of clusters was achieved. That is, it not achieved proper contingency pass to productive nodes, which caused paralysis in client services.

This article is divided into 4 sections apart from the introduction. Section 2 presents the theoretical framework. Section 3 presents the methodology used. Section 4 describes results obtained. Finally, the conclusions are presented.

2   Background

2.1   Monitoring of Infrastructure

The monitoring of an infrastructure is the process by which is brought into observation hardware or software, to ensure that it complies with the settings, key activities and established conditions [9]. This is to ensure the performance, maintain the safety and sustainability of the service or infrastructure [9]. Monitoring infrastructure enables the proper functioning and performance of the system [8]. There are several types of monitoring for ICT infrastructure, among them which we can mention: (1) Active or passive: this form of monitoring allows you to identify situations in the infrastructure from a query direct (Active) or signals sent by the agent (Passive), (2) Reactive or preventive: Reactive is the one that takes actions when there is an impact, and change a preventive analyze to a set of situations to predict a problem, and (3) measurement continues: is in constant monitoring of the hardware to ensure the correct operation of the hardware. A good monitoring service should combine different types of monitoring referred to cover all the possibilities [9]. The monitoring systems are configured to send us notification emails in case of triggering conditions for any of the monitored attributes or in the case of no data being received in a period established [6].

2.2 Monitoring Systems

Monitoring systems are software tools that utilize a collection of parameters that help us to perform an analysis of the status of networks services or servers we wish to monitor. Zabbix open-source monitoring tool provides proactive monitoring of the infrastructure [10]. It is considered as a popular tool that can be integrated with the cloud to provide monitoring services [8] of different resources (system resources, network, sensors, databases and applications) [11]. Zabbix uses a central server as a platform, from which it receives the information of the monitored resources through the execution of scripts. It can be customized and add functionalities that can display specific information in a way friendly to the user [12]. Nagios is another tool which offers at their core the possibility of building plug-ins to perform the monitoring of particular elements. Both monitoring tools are cross-platform, open source and provide support. Among the recommendations on criteria for comparing tools of monitoring are mentioned: (1) the different methods of alert (SMS, email), (2) the customization of the tool, (3) the operating systems that it supports, and integration with the systems help desk which allows integrating the monitoring system with the error-resolution process [13]. In this research we have used Zabbix [10] because it has features of customization, which facilitates the possibility of developing in-house scripts that may be compatible with the operating system of the nodes in the cluster.

Zabbix has been introduced to eliminate the problems by which the staff immediately monitor and acknowledge the impaired devices through the application not such traditional means as telephone call [14] and makes it possible to report the network devices status. Zabbix has been used (1) to monitor the status of resource usage from the server [15], (2) to report the power consumption [7], (3) to send notification emails in case of triggering conditions for any of the monitored attributes of a system of networked nest boxes that were created to study nesting birds in urban environments [6], (4) to increase the efficiency of the work carried out by the network administrators trough the real-time representation of the results in a graphical manner [16], or to manage in automatic way and real-time monitoring of a large network of a university [16].

2.3 Cluster

The cluster of servers is a set of servers that are reported to be connected to provide services [17]. A cluster is a group of independent computers that work together to increase the availability and scalability of services and applications. If one or more of the cluster nodes fail, other nodes begin to provide service, which allows users to experience a minimum of interruptions in the service.

This is achieved thanks to the interconnection of different computers through the network [18]. The clusters can be classified into High-availability that consists of moving the resources from one cluster node (IP, service, file system, among others) dropped to another node [19] or (2) Load balancing that is responsible for distributing all incoming requests between different nodes, which are used as a platform for a service application or database in such a way that if one of the nodes goes down, the other nodes will keep the service available, and distribution is used typically for environments of web-hosting [20]. The types of availability solutions based on the response time to failure are classified into Stand-alone, enhanced stand-alone, High-availability clustering and Fault-tolerant [21]. The solution Stand-alone takes days for the service to be available again, besides the data that will be the last backup performed (2) Enhanced stand-alone takes hours for which the service is available in the cluster contingency, and the data available will be that of the last transaction, that is to say, even before the primary node fails, (3) High-availability takes seconds to move resources from one node to another, that is to say, the service is available in seconds, in case the primary node fails, and (4) Fault tolerance: the service is available all the time since the time of unavailability of service is zero (in this type of cluster does not lose data).

3   Methodology

PPDIOO Cisco methodology [22] was used in the present applied research. It was adapted to the project needs. The PPDIOO Cisco methodology has the following phases: Prepare, Plan, Design, Implement, Operate, and Optimize (See Fig. 1). Below, each of the phases of the methodology is described.

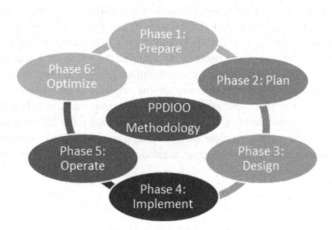

Fig. 1. PPDIOO Cisco methodology

3.1   Phase 1: Prepare

In this phase, information is collected, a summary of the origin of the project, the objectives, benefits and characteristics of the technology used for the project. Likewise, to collect the functional requirements and non-functional client. Also, in this phase we identify the cluster resources that will be monitored. Table 1 presents the roles, the tools, the activities and the deliverables of this phase.

Table 1. Activities and products of phase Prepare

Roles	Unix Specialist, Project manager, Client, and end-user
Tools	Word processor
Activities	Deliverables
Document the introduction to the project	Project Document
Describe the origin of the project	
Define the project goals	
Define the project organization structure	
Describe the benefits of the project	
Describe the characteristics of the technology used	
To make the document of the cluster resources	Document resource of the cluster
To make the document of customer requirements	Customer requirements

3.2 Phase 2: Plan

In this phase, the initials requirements are identified in function of goals. Also, the installations, users' needs and all main elements (hardware and software). In Table 2 we present the roles, the tools, the activities and the deliverables of this phase.

Table 2. Activities and products of phase Plan.

Roles	Unix specialist and Project manager
Tool	Word processor and spreadsheets
Inputs	Document presentation of the project, Document resource of the cluster and Customer requirements
Activities	Deliverables
Identify number of nodes to be monitored	Document with characteristics of monitoring system
Determine hardware and software requirements	
Determine the number of devices to be used in the project	
Identify the main system services, roles and other necessary services	
Identify the users that will benefit from the system	
Summary of the clusters to monitor	
Determine the time required for each phase of the project	Project schedule and cost estimation
Determine the duration of the entire project	
Determine the approximate cost of the entire project	

3.3   Phase 3: Design

In this phase, you integrate devices without disrupting the existing network or creating points of vulnerability. The network is illustrated with images and maps using software to assist in the topology design of the project. In Table 3 we present the roles, the tools, the activities and the deliverables of this phase.

Table 3. Activities and products of phase Design.

Roles	Unix specialist
Tools	Creately
Inputs	Document with characteristics of system
Activities	Deliverables
Diagram of the current state of the monitoring architecture cluster	Architecture design of current project
Diagram with the architecture after project implementation	Design of proposed architecture

3.4   Phase 4: Implement

In this phase, we incorporate additional components in accordance with the design specifications. Table 4 presents the roles, the tools, the activities and the deliverables of this phase.

Table 4. Activities and products of phase Implement.

Roles	Unix Specialist, VMware specialist and Networking specialist
Tools	Zabbix, spreadsheets, Sublime and Photoshop
Inputs	Architecture design proposal and Requirements of client
Activities	Deliverables
Setting server monitoring	Document server configuration Server monitoring
Configuration of the functionalities of the monitoring system	Document with the functionalities of the monitoring system
Setting events	Monitoring system the first version

3.5   Phase 5: Operate

In this phase, it performs the fault detection and the performance monitoring of the system, in addition to providing the initial data for the optimization phase. In the Table 5 presents the roles, the tools, the activities and the deliverables of this phase.

Table 5. Activities and products of phase Operate

Roles	Unix Specialist and Specialist in cluster
Tools	Zabbix and spreadsheets
Inputs	Document with the functionalities of the monitoring system Monitoring system the first version
Activities	Deliverables
Make backup copies of the software	SQL file of the database, backup scripts
Test the operation of the monitoring system	Report of tests of the system

3.6 Phase 6: Optimize

In this phase, improvements are made to the system in order to comply with the objectives of the project. In Table 6 we present the roles, the tools, the activities and the deliverables of this phase.

Table 6. Activities and products of phase Optimize

Roles	Unix specialists
Tools	Zabbix and Sublime
Inputs	Report of tests of the system Document with the functionalities of the monitoring system
Activities	Deliverables
Performs the necessary corrections in the monitoring system	Monitoring system User manual

4 Results

Table 7 present a brief summary of the project, the objectives, benefits and characteristics of the technology. Below, we present the results of applying the methodology and implement the system of monitoring of clusters. The monitoring of servers made by the company contains the following architecture:

- A central server monitoring, which acts as a control point for alerts received from agents, and collects performance data and availability.
- A server portal is used by operators to access the monitoring system. This server retrieves data from the central server monitoring in response to the actions of the operator and returns the data to the presentation.

The architecture proposed in this research includes adding a server to Zabbix central to the monitoring architecture of the company. Which will collect and analyze the information from the event cluster, as well as the resources of each node.

Table 7. Summary of the project.

Item	Description
Introduction	The project consists of implementing a web system that allows the monitoring of clusters
Problem	There is not a correct pass to the contingency of the productive nodes of the cluster, this due to the lack of knowledge of the state of the resources of the cluster and its events
General goal	Reinforce the monitoring of clusters and customize the monitoring system
Specific goals	Increase the amount of monitored resources that are part of the clusters. Increase the number of monitoring rules recorded for the detection and analysis of cluster events. Implement a web system that allows monitoring of events, nodes and cluster resources
Roles	Project manager (1), Unix specialist (2), Vmware specialist (1), and Networking specialist (1)
Benefits	Cluster resource monitoring to ensure the correct pass to contingency. Reduction of the time spent in previous validations for cluster maintenance
Technology characteristics	Software: ZABBIX (customized and funtionalities added) Hardware: Server with operative system Red Hat 7.2
Resource groups	File systems, Raw devices: discos, Volume groups, IP addresses, NFS shares, Applications, Workload partitions (WPARs)

This server will allow the specialists to Unix viewing the status of the cluster through a web interface. Likewise, it incorporates the use of the service of a mail server, owns the company, which will send the necessary notifications to the specialists, Unix responsible for managing the cluster. Figure 2 presents the proposed architecture.

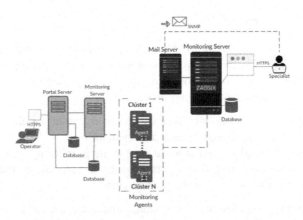

Fig. 2. The architecture of the proposed solution

4.1   Increase in Monitored Resources

With the implementation of the monitoring system we're able to increase the number of monitored resources of the cluster, due to that initially only monitored three rules and three resources in the cluster. Table 8 shows that the resources and rules monitored has increased (see Table 8). This is achieved by monitoring a more comprehensive and proactive approach, which allow them to anticipate failures that are likely to have the clusters.

Table 8. Monitoring of resources

Parameter	Before	After
Amount of monitored resources that are part of the clusters	3	14
Number of monitoring rules for the detection and analysis of cluster events	3	17
List of clusters with future availability problems	0	8
Web system that allows monitoring of events, nodes and cluster resources	0	1

4.2   Display the Status of the Cluster

The monitoring system must display the list of clusters for each client. In Fig. 3 we present the status of the nodes in the client, which consists of 6 cluster with two nodes each. Likewise, the system allows you to select each cluster and routed to another screen for verification in more detail.

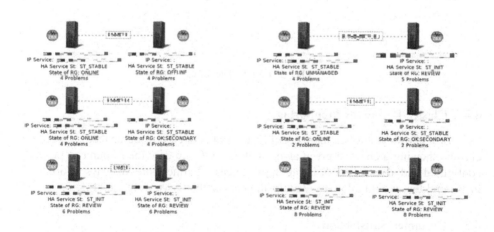

Fig. 3. Status of the nodes in the client

4.3   Monitoring Rules for Cluster Events

We managed to set up 16 new rules of monitoring associated with the detection of cluster events, because of that, the logs are configured in the monitoring system of the company are not oriented to the monitoring of events in the cluster, but the monitoring of events from the operating system of the servers, and therefore of the nodes in the cluster. The rules of monitoring events that were defined through the triggers. Trigger severity defines how important a trigger is. Zabbix supports the following trigger severities which are: Not classified: Unknown severity (grey), Information: for information purposes (light blue), Warning: be warned (yellow), Average: average problem (orange), High: something important has happened (light red), Disaster: Financial losses, etc. (red). Figure 4 shows the monitoring rules and the expression associated to trigger.

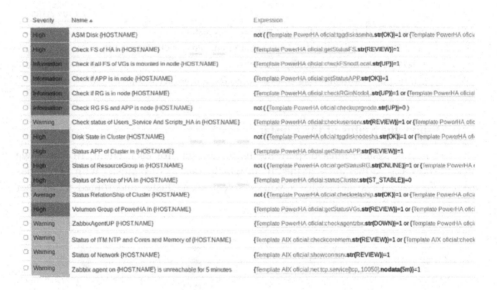

Fig. 4. Events detection.

4.4   Cluster Resources Monitoring

This view allows you to see more detail on the status of each resource that is part of the cluster, showing a checked green along with an "OK" in case the resource is properly configured and showing an X in the opposite case. Furthermore, displays information in general that would allow to identify the cluster that is viewing (see Fig. 5).

4.5   Customer Satisfaction

In order to know the customer satisfaction a questionnaire that included 8 items was designed. A customer satisfaction survey was carried out with the purpose of knowing their level of satisfaction and collecting some opinions and recommendations on the system monitoring. Table 9 shows the questions y answers.

As part of the continuous improvement is considered the suggestion of a user, which requested to send a weekly report to the email, option which was implemented.

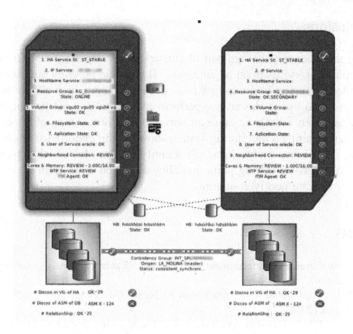

Fig. 5. Status of the cluster resources.

Table 9. Monitoring of resources

Question	Answer
Do you think the monitoring system is intuitive and easy to use	Yes = 100%, No = 0%
How satisfied are you with the new cluster monitoring system?	Very satisfied = 63,6%, Satisfied = 36,4%, Dissatisfied = 0%
Does the system have access authentication	Yes = 100%, No = 0%
Is the monitoring system available 24 h a day?	Yes = 100%, No = 0%
Do you consider the graphic interface of the monitoring system of clusters?	Yes = 81,8%, A little = 18, 2%, No = 0%
Have you received notifications of the status of clusters by the system?	Yes = 72,7%, No yet = 27,3%
Have you received a user manual to use the new monitoring system?	Yes = 100%, No = 0%
What comments do you have regarding the monitoring system? What do you think can be improved?	Very good; system must send a weekly report to the email; good solution because it saves time

5   Conclusions

Monitoring the resources that are part of clusters, is an important activity to maintain the availability of services in an organization. This article presents the implementation of the monitoring system to set up new rules of monitoring, which allows the detection of events generated by the cluster, and perform an analysis that notifies the staff in case of any detected events that may cause an impact on the functioning of the clusters. The results of implementing a monitoring system show (1) the importance in the continuity of services, (2) having early alerts, (3) identifying the clusters that present future problems, (4) to have a starting point that allows later on to perform the correction of the resources that are poorly configured.

References

1. https://www.theseus.fi/bitstream/handle/10024/94415/Bachelor_Thesis-_Anatolii_Shokhin.pdf?sequence=1
2. Cisco Networking Academy: Introduction to Networks: Companion Guide. Cisco Press, United States (2014)
3. Koetsier, J.: The high cost of server downtime (infographic). https://venturebeat.com/2012/11/14/the-high-cost-of-server-downtime-infographic/
4. Hernández, A.: Academia. https://www.academia.edu/31661093/Ciclo_de_vida_Cisco
5. Statista: Frequency of server failure based on the age of the server (per year). https://www.statista.com/statistics/430769/annual-failure-rates-of-servers/
6. Kubizňák, P., Hochachka, W., Vlastimil, O., Kotek, T., Kuchař, J., Klapetek, V., Hradcová, K., Růžička, J., Zárybnická, M.: Designing network-connected systems for ecological research and education. Ecosphere **10**(6), 1–25 (2019)
7. Kavanagh, R., Djemame, K.: Rapid and accurate energy models through calibration with IPMI and RAPL. Concurr. Comput.: Pract. Exp. **31**(13), 1–21 (2019)
8. Aparna, D., Goel, A., Gupta, S.: Analysis of infrastructure monitoring requirements for OpenStack Nova. Proc. Comput. Sci. **54**, 127–136 (2014)
9. Llorens, J.: Gerencia de Servicios de TI (2017). http://gsti.yolasite.com/
10. http://www.zabbix.com/features.php
11. Fatema, K., Emeakaroha, V., Healy, P., Morrison, J., Lynn, T.: A survey of cloud monitoring tools: taxonomy, capabilities and objectives. J. Parallel Distrib. Comput. **74**(10), 2918–2933 (2014)
12. Dalle, A., Kewan, S.: Zabbix Network Monitoring Essentials. Packt Publishing, Birmingham (2015)
13. Hernantes, J., Gallardo, G., Serrano, N.: IT infraestructure monitoring tools. IEEE Softw. **32**(4), 80–93 (2015)
14. Zabbix solutions. https://www.zabbix.com. Accessed 10 Sept 2017
15. Pacharaphak, A.: ZABBIX monitoring alert and reply with LINE Application. Master of Science, Mahanakorn University of Technology, Bangkok (2016)
16. Petruti, C., Puiu, B., Ivanciu, I., Dobrota, V.: Automatic management solution in cloud using NtopNG and Zabbix. In: 17th RoEduNet Conference: Networking in Education and Research (RoEduNet), Cluj-Napoca, pp. 1–6. IEEE (2018)
17. Purnawansyah, R.: Cluster computing analysis based on beowulf architecture. Int. J. Comput. Inform. **1**, 9–16 (2016)

18. Supercomputing Facility for Bioinformatics & Computational Biology, Supercomputing Facility for Bioinformatics & Computational Biology (2017). http://www.scfbio-iitd.res.in/doc/clustering.pdf
19. Fernández, C.: Clúster de Alta disponibilidad en Promox VE 4.2 con balanceador de carga e instalación de Pydio sobre LXC. PROMOX (2016)
20. Werstein, P., Situ, H., Huang, Z.: University of Otago. http://www.cs.otago.ac.nz/staffpriv/hzy/papers/pdcat07.pdf
21. Quintero, D., Baeta, S., Bodily, S., Buehler B., Cervantes, P., He, B., Huica, M., Knight, H.: IBM PowerHA SystemMirror V7.2 for IBM AIX Updates. IBM (2016)
22. Abdelsalam, E., Salem, K., Safieldin, A.: Development PPDIOO methodology to be compatible. Int. J. Sci. Technol. 15, 4–14 (2016)

A Data Driven Platform for Improving Performance Assessment of Software Defined Storage Solutions

Silvana De Gyves Avila(✉), Patricia Ortegon Cano,
Ayrton Mondragon Mejia, Ismael Solis Moreno,
and Arianne Navarro Lepe

IBM, Mexico Software Lab, Carretera al Castillo 2200, 45680 El Salto,
Jalisco, Mexico
{silvana.degyves,patricia.ortegon,ayrton.mondragon1,
arianne.navarro}@ibm.com, ismael@mxl.ibm.com

Abstract. Performance is one of the most important dimensions to consider for software quality. It is normally used as the principal competitive advantage offered between similar solutions. Ensuring certain performance levels is not trivial for software companies, since it requires exhaustive test runs and massive data analysis. Normally, companies rely on commercial analytics packages to produce performance insights. However, these are for general purposes and require significant effort to be adapted to particular data requirements use cases. In this paper we introduce DDP, a highly scalable data driven performance platform to analyze and exploit performance data for software defined storage solutions. DDP employs big data and analytics technology to collect, store and process performance data in an efficient and integrated way. We have demonstrated the successful application of DDP for Spectrum Scale, a software defined storage solution, where we have been able to implement performance regression data analysis to validate the performance consistency of new produced builds.

Keywords: Software quality · Performance · Big data · Data science · Software process

1 Introduction

Performance is considered one of the most important no-functional dimensions in any software development process [1]. Maintaining and exceeding performance levels, as point of competitive advantage, is fundamental for customer satisfaction and marketing goals. However, ensuring that a software component delivers the expected performance has become a complex and challenging task along with the complexity of the environments where they are intended to run [2]. Cyber-physical systems have evolved from stand-alone servers to massive distributed environments, composed of several clusters and complex workloads interacting among them. Software companies evaluating a subset of representative or realistic scenarios, require advanced tools not only to automate the execution of performance tests, but also to deal with the large amount of data produced during the evaluation process [1]. Performance data is complex itself; it

© Springer Nature Switzerland AG 2020
J. Mejia et al. (Eds.): CIMPS 2019, AISC 1071, pp. 266–275, 2020.
https://doi.org/10.1007/978-3-030-33547-2_20

comes from different sources i.e. logs, benchmarks and configuration files, with different levels of structuring. Therefore, analyzing it and producing insights, require a stack of technology and a significant effort to take the performance data through all its life-cycle. Commonly, statistical packages such as R-Studio [3], SPSS [4] and Cognos Analytics [5] are used. However, the phases in-between of collecting, storing, preprocessing and visualizing data are mostly manually carried out, creating silos where data is lost and wasted. This creates the need to rely on a holistic framework that can help performance analysts to simplify the data management and results visualization. In this paper, we introduce the Data Driven Performance Platform (DDP), an approach to integrate all the necessary technology to handle all the data coming from performance tests for Software-Defined Storage Solutions. DDP is composed of ETL and data processing engines on top of a highly-scalable data repository and interactive user interfaces. DDP has been deployed to enhance the software process of Spectrum Scale, a high-performance file management system that is being used extensively across multiple industries worldwide. DDP uses big data to produce reports related to the Spectrum Scale performance trends along with dashboards to visually support the decision making during the development and testing processes. Our successful use case shows the flexibility of DDP to handle a wide variety of data from performance tests, and how DDP successfully complements the Spectrum Scale development and testing processes to exceed the quality offered to customers.

The main contribution of this paper is the description of DDP, a platform to improve the performance testing process for software defined storage solutions and a successful use case where DDP applicability has been demonstrated. The rest of this paper is organized as follows: Sect. 2 provides the background, Sect. 3 describes the implemented data driven performance platform, Sect. 4 describes the successful use case, Sect. 5 provides our conclusions and future work.

2 Background

2.1 Spectrum Scale

In recent years, data science emerged as a new and important discipline. It can be viewed as an amalgamation of classical disciplines like statistics, data mining, databases, and distributed systems [6]. Organizations are creating, analyzing and keeping more data than ever before. The large amount of data to be processed requires massively parallel and distributed software running on tens, hundreds, or even thousands of computer nodes [7]. Distributed systems are the most adopted known strategy for storing, processing and analyzing big data [8].

Performance characteristics, such as response time, throughput, scalability and availability are key quality attributes of distributed applications used to handle data analysis requirements. Data storage becomes critical in order to meet all of these requirements [9].

IBM has a robust software defined storage portfolio, and one key component of this is Spectrum Scale, a cluster file system that provides concurrent access to a single file system or set of file systems from multiple nodes [10]. This enables high performance access to this common set of data to support a scale-out solution or to provide a high availability platform. Spectrum Scale is used in many of the world's largest scientific supercomputers and commercial applications that require high-speed access to large volumes of data [11].

When a new parallel application is developed, a stage of profiling and tuning is required to evaluate its behavior with the purpose to improve the performance. Some bottlenecks in the parallel code are only discovered in running mode. Traditional performance measures for distributed systems are: latency, that describes the delay between request and completion of an operation; throughput, that denotes the number of operations that can be completed in a given period of time; and scalability, that identifies the dependency between the number of distributed system resources that can be used by a distributed application. Other relevant metrics are I/O activities and response time [9].

2.2 Data Driven Applications in Software Development

In the last years, analytics has been applied to different fields from science to business, and software engineering is not the exception. As we can see in different works [12, 13], data driven methodologies have been applied to all the phases in the software development cycle. And the role of the data scientist in a software development team is starting to be more popular [14].

Software companies are driving a relentless focus on quality and software testing, we can say that they are facing disruption processes at the moment. The most common and repetitive tasks are been replaced by automated testing frameworks more frequently and new automation frameworks are been created [15].

A data driven approach in testing works as follows: test inputs are stored in a database, and a test driver takes multiple sets of input data and executes test cases. When test inputs are added or removed in the database, tests are not affected. A review of different works with this approach, focused on functional requirements testing can be found in [16].

When we are verifying the functional requirements of any application, frequently, the problems can be traced back to the code or design. But for the non-functional requirements, where performance plays a key role, finding the problem can be more challenging because involves different aspects outside the implementation. For example, network and hardware configurations that require the involvement of different experts in different areas.

During the development cycle, performance variations occur several times, but performance testing is the tool to avoid the impact that those negative variations could have on the next product build or release [17]. Developers need to detect performance regressions as early as possible to reduce their negative impact and fixing cost. Thus, there is a need to integrate this kind of testing into the development cycle and not leave it at the end.

Performance evaluation is usually expensive to conduct because it requires frequent executions that run from hours to days. Furthermore, the amount of metrics and data generated are often complex and produced in large volumes.

Most of the recent work that we can find in performance testing automation is focused in the analytics of the collected data. In [18], Shang proposed a model-based approach to automate the analysis of performance metrics, building regression models to find performance regressions. Foo [19], presents another approach based in the impact of different environments where performance tests are executed, using statistical models. In [20] a machine learning approach is used to interpret big data benchmark performance data.

We can also find some works targeting other steps of the performance testing process. In [17], Mostafa implements a method to prioritize performance tests cases, in order to allow more executions in earlier stages of development. Kroß [21], presents a DevOps approach for performance testing, proposing an architecture to achieve this goal. Okanović [22], presents a tool to create performance analysis reports based in concerns.

Collecting, analyzing, modeling and visualizing are part of the performance test process. However, none of the existing solutions provide the integration of all the performance evaluation phases. Tools that allow the analysis of all this data and present it to the different stakeholders, in a straightforward way, are highly required.

The different approaches described in this section, target one or two of those steps, while the others are covered using some commercial tools that are not always ad-hoc for the system under test or are done manually. Integrating all the different solutions can be difficult to accomplish and could lead to data loss. There is still a need for tools that integrate all the stages of the performance evaluation process in one. In the following sections we describe DDP, a platform to analyze and exploit performance data across all the evaluation phases.

3 DDP Platform Description

The DDP platform provides users with a centralized repository that allows them to store, organize, process and visualize performance results, into and from a single source. This addresses the issue of consuming information corresponding to performance data (benchmarks results, workload data, system profiles, hardware configuration and logs), that has diverse origins and formats, and is not explicitly related.

The components that integrate the DDP platform architecture are presented in Fig. 1 and described as follows:

- **ETL (Extract, Transform and Load).** Module that oversees reading, processing and storing key data obtained from diverse performance data files. Each data file includes relevant information that needs to be filtered and structured before being stored. This eases the numerical and textual analysis performed by the other modules of the platform. This module is implemented in Python. For each of the different benchmarks executed during the performance evaluation, exists a parser that uses regular expressions (using python's re library) to extract all relevant data. Obtained values are processed and uploaded to the corresponding tables using the defined Cassandra models.

- **Scalable Data Storage.** Repository where performance data and files are stored in a consolidated structure. It is comprised of two non-SQL Databases, Cassandra [23] and MongoDB [24]. Structured data is stored in different Cassandra tables, while complete data files are stored in MongoDB documents. Structured information will be used for analysis, whereas documents, for consultation. In some cases, it could be necessary that a human expert reviews specific details contained in those files for provenance tracking.
- **Scalable Data Processing.** Module that processes performance data stored in the Cassandra Database. Data processing is executed using Scala jobs [25] and a Spark cluster [26]. To keep response time small when consulting reports and dashboards, all results obtained by the processing module, are also stored in Cassandra structures.
- **Machine Learning and Statistics.** Set of algorithms and heuristics used to determine performance regressions, configuration profiles and all information to be displayed in the user interface. This module is implemented in Python. Further details in terms of this module cannot be included due to a non-disclosure agreement.
- **Interactive User Interface.** Module that includes all the elements required to not only visualize data, but to import the files into the system's repository and coordinate the execution of the other system modules. It is implemented using Django, with Python and Bootstrap.

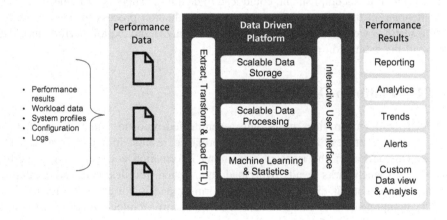

Fig. 1. Data Driven Performance Platform architecture.

This platform is oriented to be used by performance analysts and developers. Being the performance analysts those that execute diverse performance-oriented benchmarks and upload all the collected data to the platform. On the other hand, developers are more focused on the performance results presented by the tool. Information displayed includes reports, analytics, trends and alerts that can be used in the decision-making process while developing further components of software systems. The provided performance results display the behavior of newer software versions, as compared to their previous ones. By having access to analytics and trends, developers are capable to assess the quality of the product.

To validate the benefits provided by the DPP platform, it has been tested with a use case, which is described in the following section.

4   Successful Use Case

The DDP platform has been deployed in a development environment to cover the data requirements for the performance regression evaluation of Spectrum Scale. The development team performs regression evaluations to validate that new releases maintain or improve the performance in comparison to previous releases. The performance regression assessment is composed of several test cases for data and metadata operations, where standard benchmarks are used to stress the systems. These benchmarks include IOR [27], MDTEST [28] and IOZONE [29]. Every test case has variations and the main goal is to create writing and reading workloads to evaluate the performance of the different components of the Spectrum Scale configuration. During the execution of the test cases, several data files are created, including benchmark results, file system logs, configuration files and resource monitoring data. Then, all this data needs to be collected, processed and presented, to show the performance trend to the development team decision makers. The data life-cycle of this process is complex, since it requires dealing with large data sources with different levels of structuring, and needs a considerable amount of preprocessing for blending, summarizing and visualizing. To face this complexity, DDP has been deployed in three major components, as illustrated in Fig. 2. First, the application layer hosts all the front-end Web interfaces that are composed of interactive dashboards to present the results to the users. Additionally, the application layer also hosts all the backend algorithms for ETLing, data blending and summarizing. The big data storage integrates Cassandra databases with MongoDB collections to aggregate all the preprocessed data, and to store non-structured data for a highly flexible data collection. This deployment also allows to easily scale in and out the storage capacity according to the operational requirements. Finally, the processing engine integrated by Spark with Scala connects directly to the big data storage, processes the data and stores back the results to be pulled by the Web interfaces. This processing engine is also highly scalable to allow efficient processing times depending on the workload size.

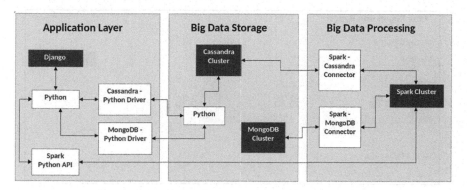

Fig. 2. Data Driven Performance deployment overview.

This deployment enhances significantly the performance evaluation. Once the data has been generated, the performance analyst using the interactive interfaces can in few clicks upload all the data. Internally, the ETL modules preprocess the data, transform it into the appropriate structures and load them into the big data storage. Here, the data is available to the processing engine for exploitation. Hence, in the case of performance regression assessment, the analyst can select the release versions to compare and obtain the expected results. This process enhancement makes transparent all the data handling complexity, reducing the processing times and making the data available for further analysis beyond the performance regression assessment.

The following figures illustrate some interfaces of the application. Figure 3 shows the interface where the analyst uploads files corresponding to regression tests. Those files populate the database used to generate the regression reports. All the data in the interfaces illustrated in this paper are examples and do not represent results from real performance evaluation.

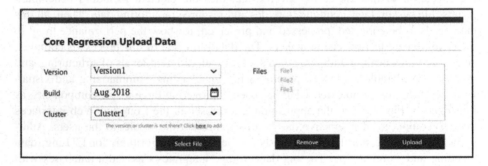

Fig. 3. Interface to upload regression data.

In Fig. 4 is depicted the "Core Regression Report", here the analyst can observe if there has been a performance improvement or regression, by comparing two different versions. By clicking on the "Setup Details" button, the analyst obtains the cluster configuration that corresponds to the set up for this regression test.

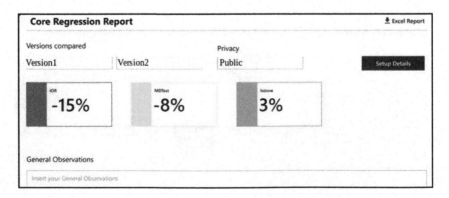

Fig. 4. Regression report.

Based on this report, developers determine if there are further actions needed before releasing the new version of the product, due to any regression presented during the performance tests.

5 Conclusion and Future Work

Performance evaluation represents a complex task due to the variety in software, hardware, network and data characteristics that can be integrated into today's systems. A data driven approach is emerging in several organizations that recognize the importance of obtaining value from the data as a competitive advantage and differentiation in the business market, and to make better decisions that improve the quality of the products.

This paper described a platform based on a data driven approach as a proposal to transform the performance testing process. This platform centralizes, analyzes and visualizes the performance data produced during the software development cycle. The architecture and technology stack for the Data Driven Performance platform was discussed, as well how this platform transforms the data coming from different sources and formats, i.e. benchmarks, logs, hardware and software configuration; and displays the result over interactive interfaces. Finally, a successful use case was described, where the deployment model was explained, and the way on how DDP is enhancing a real software process was discussed.

Some of the key conclusions that we can outline from this work can be summarized as follows:

- Data Driven is a key tool to improve software processes. It can help to derive more insights on software quality than just running functional testing. Trends, alerts, predictive and performance models can be derived from efficient data exploitation.
- A platform like DDP accelerates our journey to the AI [30]. DDP allows us to collect and centralize the data of every type from different sources. Then, transform it into data simple and accessible, organized into data models that give great sense and readiness for handling. Data is now in the analysis phase, helping developers to make faster and better decisions on regards the quality of the product.
- The platform automates the data upload and report generation. In the use case is described how the analyst can visualize if there was a performance improvement or regression just by selecting the versions of interest. There is a time saving implicit that could be re-invested in the root cause analysis.
- A data driven approach with a centralized repository allows to exploit historical data and identify trends within any period of time.
- A data driven approach with a comprehensive data visualization reduces the expert dependency to read and understand the performance evaluations.

Currently the proposed platform aims to automate the Performance Analyst activities and improve the process in the software development. Future work is being focused on exploiting the performance data in different ways, including cause-effect analysis, development of predictive models and cognitive applications that drive us to new and better performance models, and improvements in areas that are been tuned by experience and intuition rather than by deep data analysis.

References

1. John, L.K., Eeckhout, L.: Performance Evaluation and Benchmarking. CRC Press, Boca Raton (2018)
2. Aleti, A., Trubiani, C., van Hoorn, A., Jamshidi, P.: An efficient method for uncertainty propagation in robust software performance estimation. J. Syst. Softw. **138**, 222–235 (2018)
3. Dessau, R.B., Pipper, C.B.: "R"– project for statistical computing. Ugeskr. Laeger **170**(5), 328–330 (2008)
4. Aldrich, J.O.: Using IBM® SPSS® Statistics: An Interactive Hands-on Approach. Sage Publications, Thousand Oaks (2018)
5. Huijgens, H., Spadini, D., Stevens, D., Visser, N., van Deursen, A.: Software analytics in continuous delivery: a case study on success factors. In: Proceedings of the 12th ACM/IEEE International Symposium on Empirical Software Engineering and Measurement, p. 25. ACM (2018)
6. Van Der Aalst, W.: Process Mining: Data Science in Action, pp. 3–23. Springer, Heidelberg (2016). https://doi.org/10.1007/978-3-662-49851-4
7. Provost, F., Fawcett, T.: Data science and its relationship to big data and data-driven decision making. Big Data **1**(1), 51–59 (2013)
8. Roscigno, G.: The Role of Distributed Computing in Big Data Science: Case Studies in Forensics and Bioinformatics (2016)
9. Denaro, G., Polini, A., Emmerich, W.: Performance testing of distributed component architectures. In: Beydeda, S., Gruhn, V. (eds.) Testing Commercial-off-the-Shelf Components and Systems, pp. 293–314. Springer, Heidelberg (2015)
10. Coyne, L. et al.: IBM Software-Defined Storage Guide. IBM Redbooks (2018)
11. Quintero, D. et. al.: IBM Spectrum Scale. IBM Redbooks (2015)
12. Menzies, T., Zimmermann, T.: Software Analytics: So What? IEEE Softw. **30**(4), 31–37 (2013)
13. Mockus, A.: Engineering big data solutions. In Proceedings of the on Future of Software Engineering, pp. 85–99, ACM (2014)
14. Kim, M., Zimmermann, T., DeLine, R., Begel, A.: The emerging role of data scientists on software development teams. In: Proceedings of the 38th International Conference on Software Engineering, pp. 96–107. ACM (2016)
15. Yandrapally, R., Thummalapenta, S., Sinha, S., Chandra, S.: Robust test automation using contextual clues. In: Proceedings of the 2014 International Symposium on Software Testing and Analysis, pp. 304–314. ACM (2014)
16. Anbunathan, R., Basu, A.: Data driven architecture based automated test generation for Android mobile. In: 2015 IEEE International Conference on Computational Intelligence and Computing Research (ICCIC), pp. 1–5. IEEE (2015)
17. Mostafa, S., Wang, X., Xie, T.: PerfRanker: prioritization of performance regression tests for collection-intensive software. In: Proceedings of the 26th ACM SIGSOFT International Symposium on Software Testing and Analysis, pp. 23–34. ACM (2017)

18. Shang, W., Hassan, A.E., Nasser, M., Flora, P.: Automated detection of performance regressions using regression models on clustered performance counters. In: Proceedings of the 6th ACM/SPEC International Conference on Performance Engineering, pp. 15–26. ACM (2015)
19. Foo, K.C., Jian, Z.M., Adams, B., Hassan, A.E., Zou, Y., Flora, P.: An industrial case study on the automated detection of performance regressions in heterogeneous environments. In: Proceedings of the 37th International Conference on Software Engineering, vol. 2, pp. 159–168. IEEE (2015)
20. Berral, J.L., Poggi, N., Carrera, D., Call, A., Reinauer, R., Green, D.: ALOJA: a framework for benchmarking and predictive analytics in Hadoop deployments. IEEE Trans. Emerg. Top. Comput. **5**(4), 480–493 (2017)
21. Kroß, J., Willnecker, F., Zwickl, T., Krcmar, H.: PET: continuous performance evaluation tool. In: Proceedings of the 2nd International Workshop on Quality-Aware DevOps, pp. 42–43. ACM (2016)
22. Okanović, D., van Hoorn, A., Zorn, C., Beck, F., Ferme, V., Walter, J.: Concern-driven reporting of software performance analysis results. In: Companion of the 2019 ACM/SPEC International Conference on Performance Engineering, pp. 1–4. ACM (2019)
23. The Apache Software Foundation: What is Cassandra? http://cassandra.apache.org/
24. MongoDB, Inc.: Why MongoDB. https://www.mongodb.com/
25. École Polytechnique Fédérale: The Scala Programming Language. https://www.scala-lang.org/
26. The Apache Software Foundation: Apache Spark. https://spark.apache.org/
27. Lawrence Livermore National Laboratory: IOR Benchmark. https://github.com/LLNL/ior
28. Lawrence Livermore National Laboratory: MDTEST Benchmark. https://github.com/LLNL/mdtest
29. IOZONE org: IOZONE Filesystem Benchmark. http://www.iozone.org/
30. IBM: Accelerate the journey to AI. https://www.ibm.com/downloads/cas/MZEA2GKW

Teaching Approach for the Development of Virtual Reality Videogames

David Bonilla Carranza[1], Adriana Peña Pérez Negrón[1(✉)],
and Madeleine Contreras[2]

[1] Universidad de Guadalajara CUCEI, Blvd. Marcelino García Barragán #1421,
44430 Guadalajara, Jalisco, Mexico
jose.bcarranza@academicos.udg.mx,
adriana.pena@cucei.udg.mx
[2] Universidad de Guadalajara, Sistema de Universidad Virtual, Av. Enrique Díaz
de León no. 782, Col. Moderna, 44190 Guadalajara, Jalisco, Mexico
mgabriela@suv.udg.mx

Abstract. This proposal was developed based on the simulation subject for
undergraduate students of Computer Science Engineering, which objective is
that the students obtain general knowledge and techniques to develop videogames based on virtual reality for mobile devices. The subject covers from the
game conception to its final presentation in a functional product. Building on the
teaching experience of this subject, it was observed that students found difficulties on relating different technical aspects to develop a project of this type.
Therefore, looking to correct these deficiencies, this article describes a proposal
based on Project-Based Learning (PBL) and videogames development method,
linking material, tools, and creative knowledge to establish the teaching process
for videogames. This approach considers creative, technical and practical
components improving the students' learning experience. The students are
required to present a final project to solve a social problem based on videogames, some of which are here presented.

Keywords: Videogame development · Virtual environments simulation ·
Mobile devices · Project-based learning

1 Introduction

Virtual reality (VR) industry is constantly growing; it has become a fashionable trend.
Although it is not a tangible reality in the domestic sphere, several companies are
positioning themselves in the market and others are trying to take advantage of their
traditional domain position so they can get on board [1].

One of the main challenges for VR is being part of people's daily lives, which
represents a not easy task [2]. However, the continuous decline in prices for special
devices, particularly for immersive virtual reality like head-mounted display (HMD),
along with better computer performance is bridging the gap.

VR is intrinsically related to videogames. A 3D-graphics based videogame is VR,
defined as a technology that conveys three-dimensional graphics with users' interaction

© Springer Nature Switzerland AG 2020
J. Mejia et al. (Eds.): CIMPS 2019, AISC 1071, pp. 276–288, 2020.
https://doi.org/10.1007/978-3-030-33547-2_21

[3]. Videogames is a billion-dollar industry that keeps growing; it is software games that can be played in electronic platforms such as computers or consoles, where the emergence of smartphones and tablets represents a third category, the mobile platform.

The development of videogames requires two main areas for its creation. One is the technological area, which involves the specific definition of hardware and software as elements in the development process; while the other one is the creative area, which is the conception of the idea. The creative area involves planning, art design, and defining the narratives integrated into the game elements [4]. The whole process involves a collaborative effort between design and programming.

Although VR utilizes evolving hardware that constantly improves, the workflow remains the same, that is, concept, preproduction, production, testing, polishing, and delivery [5].

Currently, the demand for specialized professionals in VR is increasing. Therefore, it seems beneficial to start preparing students for this. Specialized institutes that focus on the development of videogames are beginning to incorporate its teaching as formal subjects for undergraduate students [1].

In this context, in this paper an approach to teaching the development of virtual reality-based videogames for mobile devices is formally presented. This approach has been applied and improved during eight half-year subjects of the computer simulation subject for undergraduate students of Computer Science Engineering. The approach is based on the videogame development methodology of Acerenza, et al. [6], and the Project-based learning instructional methodology.

2   Background

Related works about teaching approaches in software development do not specialize in the development of videogames, which might lead to misunderstandings in their study causing the lack of an adequate teaching strategy.

Although there is extensive literature on learning how to develop videogames, or even virtual reality videogames, as far as we know, there are no research papers with teaching approaches for virtual reality videogames. Claypool and Claypool [7], designed a teaching approach for the Software Engineering subject applying a video-game development process. According to them, by including videogames in the subject, the class enrollment increased, grades significantly improved, and subjective comments from students suggest a greater interest in Software Engineering over-all. Their preliminary evaluation suggests the merits of the approach. However, this approach uses videogames as a means to motivate learning and not as the main objective to understand their correlation of elements in the process of creation as a whole, considering that the development of a videogame has a technical process and a creative process [8].

2.1   Videogame Development Process

The videogame teaching process is very similar to their development process [9], see Fig. 1, where Acerenza et al. [6] described the video game development phases based on the Scrum Master and Product Owner Scrum methodology but integrating the roles of the Scrum framework. The process manages four development roles that involve the creation of a videogame: development, internal producer, client, and beta tester.

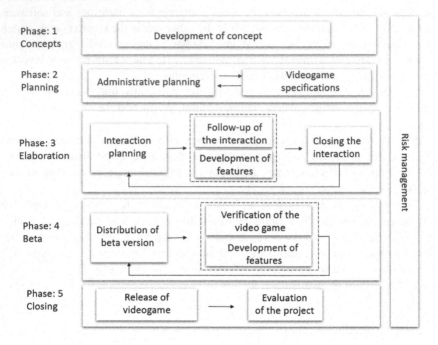

Fig. 1. Development phases for videogames taken from [6].

The development team and the client correspond directly to Scrum roles respectively [10]. The development team has the characteristics of the Scrum team, but unlike Scrum, sub-roles are defined within the team. These correspond to those commonly used in the videogame industry: programmer, graphic artist, sound artist, and game designer [11]. Based on this development process, our teaching approach provides flexibility to define life cycles and can be very well adapted to the PBL methodology.

Habitual videogames consumers are used to certain usability interaction formulas, but VR represents a new dimension. The fourth wall is broken thanks to the need for the monitor, it needs to have two cameras, one in each eye and 360-degree vision. The fourth wall refers to the metaphor of that invisible imaginary wall that is in front of a video game through which the audience sees the stories of the games. Rather, it is what separates the life of the characters from any spectator [12].

According to Rodríguez [12], for the development of VR there are 3 important challenges when using mobile devices as a platform:

1. The *movement system*, the movement of the player by the game world cannot follow the typical game rules. In this case, it is important to be cautious of not producing

unnecessary "dizziness". This dizziness effect is mainly due to the displacement of frames that do not match with those in the real world, they are also known as cyber annoyances [13].

2. The *player's camera* has to be more *natural and flexible*. It will be able to play more with the positioning of game assets, and the close exploration of environments. As a counterpart, this will force the designer to care for the quality of the nearby texture.

3. The *differences in interaction* with the environment. The gamepad is not the only control system, for example, the new tracking system of the hands radically changes the way of interacting and playing [12].

This highlights some differences for VR videogames, particularly for the mobile devices platform, implying different device issues [9]; considerations that require to be included in the teaching process. On the other hand, because the videogame teaching involves developing a videogame, the PBL seems an appropriate instructional methodology.

2.2   Project-Based Learning (PBL)

The development of software for video games should be taught not only as a tool of interest for individualized learning, but also as a support for team learning and the creation of information. PBL [14] was considered also because teamwork is proposed to carry out learning in work teams during its application [15].

Larmer et al. [16] presented their gold standard for the PBL based on 7 elements to start, develop and finish a project: a challenging problem, sustained consultation, authenticity, voice and student choice, critical reflection and review, and a public product. The use of these elements in the design of this approach guarantees a learning experience to acquire the knowledge for creating a proper project.

For a student to learn, the student must identify the difficulties and mistakes during the development process. This phase is accompanied by the basic definition that reinforces the conceptualization of the workflow in the development of projects to be able to overcome them; this is an intentional exercise [17]. According to PBL, this is self-regulation learning, which is a self-directed process through which learners transform their mental abilities into academic skills.

For PBL to support students during the learning process, two fundamental elements are used: evaluation strategies and classroom management in collaborative workgroups [15], which matches the teamwork process for videogames.

In the subject of simulation, it is sought that students acquire skills and integrate theoretical and practical knowledge and relate them to a project that might solve a social issue as a final product [14]. The PBL integrated into the videogame development process is an integral methodology where teaching expectations end on a functional product. In the next section are detailed the phases of the proposed teaching approach.

3   Teaching Process for Videogames

As mentioned, the proposed approach uses as a reference the PBL supported by the SUM adapted for the development of videogames using the structure and roles of Scrum. Also, including the Acerenza et al. [6] methodology for videogames, leads to the creation of a functional product through the continuous improvement in the process to increase its effectiveness and efficiency.

In Fig. 2 are presented the phases that comprise the integration of PBL and the Acerenza et al. [6] process. This is evaluated at the end of the cycle. During its execution students acquire skills and knowledge through the development of practices that reinforce the basic concepts, all aimed to propose how to solve a social problem using a VE videogame for mobile devices.

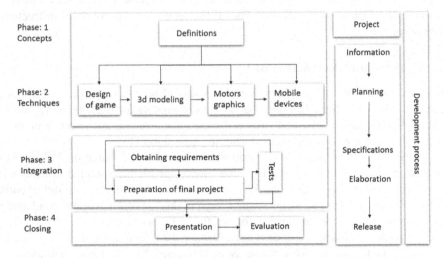

Fig. 2. Phases for the teaching process in the development of a VR videogame based on [6].

This teaching approach exposes predictable results, students must manage time resources efficiently considering probable risks during the development of the project and also managing constant interaction with their teammates [18]. When using SUM, they band together to adjust small multidisciplinary teams from three to seven members working in the same physical location or distributed, and in a short project with a duration of fewer than 6 months, looking for a high degree of participation.

First Phase Concept. In this phase, basic concepts are transmitted where workflow can be known in the development of a VR videogame for mobile devices. It includes integrating user interface design techniques [11], low polygon modeling, video game mechanics based on graphic engines. The way to evaluate this knowledge is through relevant activities, such as summaries or conceptual maps.

Second Phase Techniques. The technical phase is related to the concept phase reinforced with practices such a definition, and learning specific software [19]. At this point, each practice must have an incremental relationship so that as a whole they can later be integrated into a final project.

Third Phase Integration. It is sought that the practices developed in the previous stage serve to acquire fundamental techniques in the development of VR videogames. That is why during their realization the students are being focused on a final project. In this case, the problems to solve most have somehow a social impact that affects the environment. The integration phase consists of how the learning process is directed to the collaborative learning project based on the knowledge previously acquired.

Fourth Phase Closing. In the closing phase, feedback is received on the finished product and it is presented to the class. In the final presentation, a functional prototype is shown with special VR glasses, where the students receive constructive cues for their project.

In each of the four phases, activities that evoke a process of PBL are carried out. These are aimed to reinforce knowledge; the objective is for the student to create key parts of the project development along with evidence for evaluation. We consider this as a comprehensive way to understand the complete process involved in the videogame development project.

The teaching approach proposal is presented in Table 1. In the first column (Phase) are the phases of the teaching approach, in the second column (Software development) the stages of the software development that correspond to the phase are established, in the third column (Applications tools) the software tools with which the project is developed are named, in the fourth column (PBL elements) are the elements of the PBL from the gold standard, and finally, in the fifth column (PBL evaluation) are the elements that should be evaluated according to the PBL approach.

Table 1. Teaching approach phases and their relation linking PBL and software development.

Phase	Software development	Applications tools	PBL elements	PBL evaluation
1.- Concepts	Analysis, planning	Inkscape, Gimp	Challenging problem, sustained inquiry	People
2.- Techniques	Design	Blender,	Authenticity, student voice and choice	Processes
3.- Integration	Development	Unity engine	Reflection	Processes
4.- Closing	Release	Android, Google cardboard	Critical, revision and public product	Product

As can be observed in Table 1, for each phase from the videogame development are related the software development stages, the application tools to be teach, the PBL elements and evaluation, integrating all the elements in a single process, which takes guides the teaching process for the semester.

In such a way that for example, the first phase Concept is linked to the software development stages of Analysis and Planning, where the suggested applications tools are Inkscape and Gimp (described in detail in the next Sect. 3.1), to cover the PBL elements of Challenging, Problem, and Sustained Inquiry, which will help to evaluate People. Students at this stage determine the scope of the challenge to be solved.

3.1 Software Applications

The suggested support applications for the technical phase used in the simulation subject are the following.

For the interface elaboration vector graphics manipulation software is needed. Inkscape™ (https://inkscape.org/) is used since it allows fluidly generating game resources; it also has the characteristic of being free software.

For 3D object modeling Blender™ (https://www.blender.org/) is used. Blender generates scenarios, characters, and all the elements that will be in the environment to be developed. This software helps to create textures or material to give realism to the scene. This is also free software.

Due to its versatility, Unity™ (https://unity.com/) is used as a videogame engine by making use of a series of programming libraries that allow the design, creation, and representation of a videogame. Unity has a free software version for nonprofit developments. Unity most outstanding aspects are its graphic capabilities responsible for displaying 2D and 3D images on the screen, as well as calculating some aspects such as polygons, lighting, textures, among other features. The ability to export the game to different platforms, in this case to virtual reality environments, is one of the many reasons it is used.

The Unity graphic engine also helps in the handling of physics; this is what makes possible to apply physical approaches to videogames so that they have a real feeling in the interaction of objects with the environment. In other words, it is responsible for making the calculations for an object to simulate physical attributes such as weight, volume, acceleration, gravity, etc.

Unity also incorporates the sound engine, which is responsible for loading tracks, modifying their bit rate, removing them from playback, and synchronizing them among other things.

All the video game engines have a programming language that allows the implementing functioning characters and objects that are part of the game. Unity uses C# as a programming language to write scripts, which allows us to understand the necessary mechanics in virtual environments.

Before being able to visualize a project, a virtual reality through Goggles is required to export it to a mobile device. That is why the Android ™ tools SDK library is used to export it directly to a mobile device to test it out.

Finally, a library that is integrated into the graphics engine makes the split-screen for the vision that generates virtual reality, the Google Cardboard. Google Cardboard™ is responsible for transforming any smartphone with Android OS into a VR platform in a very accessible way using a cardboard structure as a cost-effective device that can be acquired with a guide at Google™ (https://vr.google.com/cardboard/developers/).

3.2 Phases with Projects Samples

As explained in the previous section, each software tool provides a phase of project development through specific practices. These practices are focused on teaching different concepts of the software development process of videogames based on VR.

Sample of some real practices performed by students in the computer simulation subject along with a brief description of the phase are here presented.

At the beginning of the course, the basic concepts that correspond to the first phase are taught. Here, students carry out practices focused on the following topics; videogame design, main components of VR, principles of 3D modeling and the creation of elements for the design of user interfaces. The students are required to make a concept map workflow at the end of this phase.

Once basic concepts are comprehended in the second phase students acquired technical skills using applications. Figure 3 shows a designed scenario in Blender^{TM}.

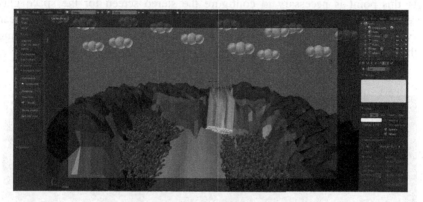

Fig. 3. Modeling example of a 3D scenario.

Subsequently, in the third phase Integration teamwork starts. The team members discuss how to treat a social problem. At this point they can define the basic resources they will need for the videogame. They translate the 3D objects to the game engine integrating the game mechanics, the user interfaces, sounds, treasures, and programming scripts to create the mobile application.

In Fig. 4 can be seen the export process to mobile dissipative. The screen is not yet divided; therefore the tests can be applied directly in a PC.

Fig. 4. First game tests before exporting it to a mobile device.

At this point is necessary to configure the stereo screen for the space glasses. Figure 5 shows the final viewpoint with divided screens. The menu creation can be observed. The input methods are programmed like for example the Bluetooth gamepad and the gyroscope, the project is still running in a PC.

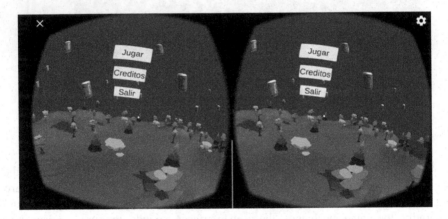

Fig. 5. Divided screen tests and input methods.

Figure 6 shows the implementation of the project. In this specific project, the pollution problem was boarded. In this phase, the game mechanics are tested indicated at the beginning of designing the game.

Fig. 6. The user interface of a working project.

Figure 7 shows the VR application running on the mobile device. This is the first test with external control, it can be seen the displacement of the character within the game level.

Fig. 7. Virtual Reality user interface tests on a mobile device.

Figure 8 shows the goggles, where the mobile device is introduced. The goggles are placed over the eyes of the user to simulate the required vision for VR. There are a wide variety of VR goggles available, the most common one is the low-cost glasses made from Google™ Cardboard.

Fig. 8. Virtual Reality cardboard glasses for mobile devices.

Finally, in the fourth phase Closing the final project is presented, Fig. 9 shows the VR interface for detecting anomalies that involve the interaction of the virtual interface. This point can be detected if the displacement is correct or if there are dizziness problems that need to be fixed. It is important to run validation and verification tests.

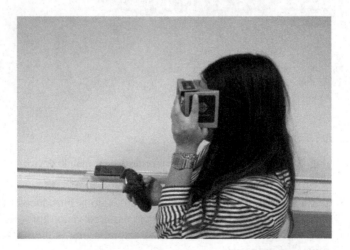

Fig. 9. Virtual Reality test of the final project.

The final product is made a functional project that requires the student to acquire the subject knowledge.

4   Conclusions and Future Work

The construction of virtual worlds is a complex task, so it is necessary to use a teaching approach that allows the student to develop focused skills to understand the complete process.

In this paper, a teaching approach to develop virtual reality-based video games for mobile devices was presented. It is important to highlight that teaching methods need to be adaptable for new technologies to come, allowing new generations to enjoy dynamic and interactive content development learning. It is worth to mention that the software proposed for this development is based on free software. This allows having constant access to the tools [7].

We consider that a learning system based on projects is an excellent way to transmit knowledge and getting feedback on a specific topic. In the future, we will try to test this teaching approach on a virtual platform environment to know its transcendence and real scopes applied in other means of knowledge transmission.

Students' general comments support the merits of the approach; they show a great interest in the virtual reality development process as a whole.

Through several semesters, we have gotten good comments from the students. Many of them refer to the subject as one in which they improved their technical and project development skills. They naturally discover the importance of knowing the workflow of a VR development on time and understanding the differences with traditional software development. It is very favorable for students to learn during the development of a project and to interact in teams, the social part help them to solve doubts.

Also, it should be noted that the projects that are carried out as a team presenting at the end of the course, seek to solve a social problem in your area. This also motivates the students to address a problem from another viewpoint. Some of the projects presented are awareness of garbage collection and separation, take precaution to the safety of the city, virtual tours designed for first-year students, taking care of forest by creating simulations of forest fires, among others.

The structure of the subject content is presented in this teaching method that reinforce the development of the project by addressing both the theoretical and practical aspects.

References

1. Cuenca, D.: Game studies: Estado del arte de los estudios sobre video juegos. Luciérnaga Rev. Virtual **10**, 13–24 (2018). https://doi.org/10.33571/revistaluciernaga.v10n19a1
2. Mora, M.A., Lopez, I., Meza, C.A., Portilla, A., Sánchez, N., Sanchez, C.R.: Metodología para el Desarrollo de Mundos Virtuales, con un Caso de Estudio: Independencia de México. vol. 10, pp. 6–8 (2014)
3. Cantero, J.I., Sidorenko, P., Herranz, J.M.: Realidad virtual, contenidos 360° y periodismo inmersivo en los medios latinoamericanos. Una revisión de su situación actual. Rev. Comun. **17**, 79–103 (2018). https://doi.org/10.26439/contratexto2018.n029.1816

4. Tschang, T.: When does an idea become an innovation? The role of individual and group creativity in videogame design. Presented at the Summer Conference, Copenhagen, Denmark, pp. 1–26 (2003)
5. Jiménez, E.M., Oktaba, H., Díaz-Barriga, F., Piattini, M., Revillagigedo-Tulais, A.M., Flores-Zarco, S.: Methodology to construct educational video games in software engineering. In: 2016 4th International Conference in Software Engineering Research and Innovation, pp. 110–114. IEEE, Puebla (2016). https://doi.org/10.1109/CONISOFT.2016.25
6. Acerenza, N., Coppes, A., Mesa, G., Viera, A., Fernandez, E., Laurenzo, T., Vallespir, D.: Una Metodología para Desarrollo de Videojuegos. In: 38° JAIIO - Simposio Argentino de Ingeniería de Software, pp. 171–176 (2009)
7. Claypool, K., Claypool, M.: Teaching software engineering through game design. In: Proceedings of the 10th annual SIGCSE Conference on Innovation and Technology in Computer Science Education, ITiCSE 2005, Monte De Caparica, Portugal, pp. 123–127 (2005). https://doi.org/10.1145/1067445.1067482
8. Chamillard, A.T.: Introductory game creation: no programming required. In: SIGCSE 2006, Houston, Texas USA, vol. 39, pp. 515–519 (2006). https://doi.org/10.1145/1121341.1121502
9. Kasurinen, J.: Games as Software: Similarities and Differences between the Implementation Projects. In: Proceedings of the 17th International Conference on Computer Systems and Technologies 2016 - CompSysTech '16. pp. 33–40. ACM Press, Palermo, Italy (2016) https://doi.org/10.1145/2983468.2983501
10. Pascarella, L., Palomba, F., Di Penta, M., Bacchelli, A.: How is video game development different from software development in open source? In: Proceedings of the 15th International Conference on Mining Software Repositories - MSR 2018. pp. 392–402. ACM Press, Gothenburg (2018) https://doi.org/10.1145/3196398.3196418
11. León, N., Eyzaguirre, S., Campos, R.: Proceso de diseño de software y proceso de diseño sonoro para un videojuego educativo. Campus 22, 27–34 (2017). https://doi.org/10.24265/campus.2017.v22n23.02
12. Rodríguez, T.: Cómo se hace un juego de realidad virtual, Xataka. https://www.xataka.com/realidad-virtual-aumentada/como-se-hace-un-juego-de-realidad-virtual. Accessed 3 May 2019
13. Cuevas, B.G., Valero, L.: Efectos secundarios tras el uso de realidad virtual inmersiva en un videojuego. Int. J. Psychol. 13, 163–178 (2013)
14. Remijan, K.W.: Project-based learning and design-focused projects to motivate secondary mathematics students. Interdiscip. J. Probl.-Based Learn. 11, 1 (2017). https://doi.org/10.7771/1541-5015.1520
15. Pérez, M.: Aprendizaje basado en proyectos colaborativos. Rev. Educ. 14, 158–180 (2008)
16. Larmer, J., Mergendoller, J., Suzie, B.: Setting the Standard for Project Based Learning - Based Learning, Assn, Virginia, EE. UU (2015)
17. Marti, J.A., Heydrich, M., Rojas, M., Hernández, A.: Aprendizaje basado en proyectos: una experiencia de innovación docente. Revista Universidad EAFIT, Colombia, vol. 46, pp. 11–21 (2010)
18. Zarraonandia, T., Ruíz, M.R., Díaz, P., Aedo, I.: Combining game designs for creating adaptive and personalized educational games. In: Presented at the the 6th European Conference on Games Based Learning, Ireland, October, pp. 559–567 (2012)
19. Dickson, P.E.: Using unity to teach game development: when you've never written a game. In: Proceedings of the 2015 ACM Conference on Innovation and Technology in Computer Science Education - ITiCSE 2015, pp. 75–80. ACM Press, Vilnius (2015). https://doi.org/10.1145/2729094.2742591

Author Index

© Springer Nature Switzerland AG 2020
J. Mejia et al. (Eds.): CIMPS 2019, AISC 1071, pp. 289–290, 2020.
https://doi.org/10.1007/978-3-030-33547-2

Printed in the United States
By Bookmasters